THE TWO RAINBOW SERPENTS TRAVELLING:

Mura track narratives from the 'Corner Country'

THE TWO RAINBOW SERPENTS TRAVELLING:

Mura track narratives from the 'Corner Country'

Jeremy Beckett and Luise Hercus

ANU

THE AUSTRALIAN NATIONAL UNIVERSITY

E PRESS

Published by ANU E Press and Aboriginal History Incorporated
Aboriginal History Monograph 18

This title is also available online at: http://epress.anu.edu.au/two_rainbow_citation.html

National Library of Australia
Cataloguing-in-Publication entry

Author: Hercus, L. A. (Luise Anna), 1926-

Title: The two rainbow serpents travelling [electronic resource] : mura track narratives from the Corner
             Country / Luise Hercus, Jeremy Beckett.

ISBN: 9781921536922 (pbk.)   9781921536939 (pdf)

Series: Aboriginal history monographs ; 18.

Subjects: Rainbow serpent.
             Aboriginal Australians--Religion.

Other Authors/Contributors:
             Beckett, Jeremy.

Dewey Number: 299.9215

Aboriginal History Incorporated

Aboriginal History is administered by an Editorial Board which is responsible for all unsigned material. Views and opinions expressed by the author are not necessarily shared by Board members.

The Committee of Management and the Editorial Board

Contacting Aboriginal History

All correspondence should be addressed to Aboriginal History, Box 2837 GPO Canberra, 2601, Australia. Sales and orders for journals and monographs, and journal subscriptions: Thelma Sims, email: Thelma.Sims@anu.edu.au, tel or fax: +61 2 6125 3269, www.aboriginalhistory.org

Aboriginal History Inc. is a part of the Australian Centre for Indigenous History, Research School of Social Sciences, The Australian National University and gratefully acknowledges the support of the History Program, RSSS and the National Centre for Indigenous Studies, Australian National University.

WARNING: Readers are notified that this publication may contain names or images of deceased persons.

ANU E Press

All correspondence should be addressed to:
ANU E Press, The Australian National University, Canberra ACT 0200, Australia
Email: anuepress@anu.edu.au, http://epress.anu.edu.au

Cover design by ANU E Press

Contents

Acknowledgments

We are deeply grateful to the main storytellers of half a century ago, George Dutton and Alf Barlow: the *Ngatyi* story and the vision of the landscape belongs primarily to them. Other important contributors of the past were Walter Newton, Hannah Quayle, George McDermott and Cecil 'Knocker' Ebsworth: they too shared with us their knowledge of the Rainbow serpents, and 'King' Frank Miller contributed unique background information on the Corner Country.

We would like to thank Badger Bates, who gave us the cover illustration: he and Sarah Martin helped and encouraged us throughout and gave us the photo of Ularara waterhole. Colin Macdonald drew all the maps. We are indebted to Isabel McBryde for her photos, and also to Pamela Macdonald, John McEntee, Simon Holdaway, Kim McCaul, John Jorgensen, Graham Hercus and Anne-Mari Siiteri-Hercus. Our thanks too go to Peter Read, Ingereth Macfarlane, Geoff Hunt and the board of Aboriginal History.

Preface

I am very pleased that Luise and Jeremy have published the Two *Ngatyi* story and appreciate the amount of work they have put into this. When I was very small my grandmother took me past Peery Lake and up to Nocoleche and Yulara Waterhole, on our way to Wanaaring to stay with family. These are all places in the Two *Ngatyi* Story, a story of great significance to the Paakantji people. We camped at Yulara waterhole, I just remember this because I was very little, and I went back there recently to take the photos for the story. My linoprint *Ngatyi Yarilana* is about the male and female *Ngatyi* at Yulara waterhole where they started their journey. This book will help the young Paakantji people learn about their culture, but it will also help people from other communities and cultures understand how the *Ngatyi* made the landscape and introduce them to our Paakantji culture.

Badger Bates

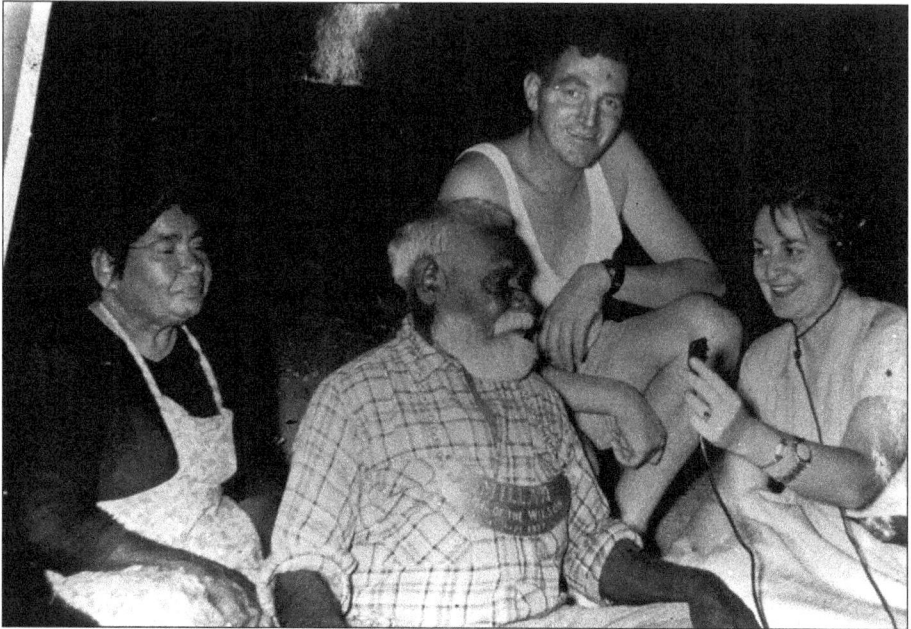

Figure 1: King Miller recording a song for a Sydney journalist in 1956. From left to right Alice Miller, King Miller, Sgt Keith Jorgenson and the Sydney journalist. Photograph courtesy of John Jorgensen.

Introduction 1

The boundaries of New South Wales, South Australia and Queensland come together at Cameron's Corner: the surrounding areas of those three states are known as 'the Corner Country'.

Both geographically and culturally the Corner Country is a very special place. It is west of the Darling and the other big rivers that come down from Queensland and flow into the Murray-Darling Basin. The Cooper Basin is not far to the north, but the only major river that flows right into the area, the Bulloo, forms a drainage basin of its own. After rains in the northern part of the Grey Range and around Quilpie the river fills up Bulloo Lake and the smaller lakes nearby. Jeremy Beckett visited Tibooburra, in the heart of the Corner Country, in company with George Dutton, the widely respected Aboriginal elder from Wilcannia in 1958. On that occasion, the most senior Tibooburra man, King Miller (Wangkumara/ Punthamara) sang a song to his visitors about Lake Bulloo, 'the big lake', after rains: this place meant so much to the people of the Corner Country. He translated some of the lines:

> Weedy water,
>
> Water lapping
>
> Wave come back
>
> Leaping up, striking the bank
>
> Water shaking
>
> Come and carry the weeds away
>
> Bulloo Lake, water
>
> Looking at it, he (a mythical Ancestor) came there.[1]

The water ultimately flows out into the big Carryapundy and Bullegree swamps, the Torowoto and numerous smaller swamps and the Bulloo overflow. These areas then spring to life with vegetation and with waterbirds. They were a major resource for Aboriginal people: for the explorers, they were the dreaded 'Mud Plains'.[2]

The nearby parts of the Grey Range have some spectacular scenery, though they only reach an altitude of 332 metres (Mt Shannon). In those rugged hills and on the stony plains there are many sites of traditional mythological significance, stone arrangements and important quarries which tell of many aspects of Aboriginal occupation. Cameron's Corner itself is in sandhill country, which continues into South Australia and has links to the Strzelecki Desert. The story of the Two Rainbow Serpents links the extreme eastern edge of the Corner Country, the Paroo at Ularada waterhole, with the far western edge, where the Flinders Ranges rise out of the sandy plain.

The Aboriginal people of the Corner Country belonged to a number of different 'tribes', who spoke languages of four different language affiliations:

1. Karnic Languages of the Lake Eyre Basin (spoken in the Corner Country by Wangkumara people from along the Warri-Warri Creek across to the Bullagree)

2. Paakantyi, Darling River Languages (spoken by Wanyiwalku/Pantyikali people from the Lake Yantara area eastwards)

3. Yarli Languages (spoken by Malyangapa 'Lake-water', Wadigali, and Yardliyawara people from the Cobham Lakes across to the Flinders Ranges)[3]

4. Karlali and the Bidjara dialect along the Bulloo (Karlali was sometimes grouped with Karnic, but recent work has shown that it was separate).[4]

Languages between these different affiliations were not mutually comprehensible, so while a Yardliyawara man could easily understand Malyangapa he would not have understood Wangkumara. The Corner Country formed a special cultural community, and these linguistic divisions were no barrier: as explained in Section 1.4 below, people were generally multilingual.

An important feature of the Aboriginal culture of the 'Corner Country', is the *mura* which corresponds to what is elsewhere in Australia called the 'dreaming track'. Though the 'classical' Central Australian concept of the Dreaming (*tjukurpa*)[5] is not part of local belief, a *mura* story is something similar. It is a narrative which details the travelling of ancestral beings also called *mura* – occasionally human but more often anthropomorphic animal – who form the country and name it as they go. The Corner Country, where both of the authors have worked, is criss-crossed by a network of such stories. It would seem that both men and women once knew a great number of *mura* stories, in greater or lesser detail. Ceremonial life ceased in the early years of the twentieth century, and the transmission of traditional knowledge began losing its appeal for the rising generations. However, in our work in the Corner, from the late 1950s to the late 1960s, we found a number of senior Aboriginal people, particularly George Dutton, Hannah Quayle, Alf Barlow and Walter Newton, who were knowledgeable in these matters, and willing to have us record them.

Figure 2: George Dutton in Broken Hill, 1957.
Photograph by Jeremy Beckett.

Figure 3: George Dutton and his son Jim at Wilcannia, 1956. Photograph by Jeremy Beckett.

Figure 4: Alf Barlow at Wilcannia, 1957.
Photograph by Jeremy Beckett.

Pastoral settlement reached the Corner Country in the 1860s.[6] By the time our *mura*-tellers were born, Aborigines were still living on their own country, not so much in their own right as by providing various services for the stations in return for food, clothing and tobacco. Yancannia, a relatively well-watered property in this generally arid region, was one of the most important areas for Aborigines, perhaps because it was the site of *Birndiwalpi* spring that figures in the story of the Two *Ngatyi* or Rainbow Serpents. From early on, the settlers transformed the landscape, particularly by sinking wells and bores at sacred waterholes and springs,[7] and this unfortunately also happened to the *Birndiwalpi* 'Brindiwilpa' spring. Nevertheless these same settlers interfered little in the Aborigines' affairs and ceremonies, not obstructing and perhaps sometimes enabling large regional ceremonial gatherings. According to our informants, these occasions attracted a number of named groups from around the Corner Country. Perhaps this mixing was a response to depopulation, but it may always have been this way – with different 'mobs' participating in one another's ceremonies, and sometimes initiating one another's boys according to their own distinctive rites.[8] Singing the *mura* was quintessentially an inter-tribal activity, which at least notionally linked peoples over considerable distances.

1.1. *Mura* stories as cultural links

'Tribal' maps now play an important part not only in land-claims, but also in much of our thinking about Aboriginal Australia. Looking at them one sees the whole of the country divided into several hundred little sections, sometimes in different colours, as in the AIATSIS map by David Horton and the map by Stephen Davis.[9] These maps fulfil an important function, but they leave us with an idea of extreme fragmentation. Aboriginal society was not like that: people interacted with their neighbours and even people further afield on a regular basis, they held joint ceremonies with other groups, not only for the purposes of initiations, but also the increase of various food sources, and sometimes simply as celebrations of ritual and song.

The *mura* stories and the long lines of song were linked to particular sites, but they also spanned a number of territories and created strong ritual links between groups sometimes quite a distance apart: great ceremonies brought people from afar. Such gatherings, as is well known, involved the exchange of goods and also marriage arrangements, which further consolidated the links.

In the dry interior everything relating to water was vital, and so stories and ritual concerning waterbirds and the legendary Water-snakes were particularly important.

Figure 5: George McDermott on the verandah of the Bourke Hospital, 1970. Photograph by Graham Hercus

George McDermott, who was Wangkumara, explained very clearly how people in the Corner Country felt about their neighbours regarding these ceremonies. He had just sung a verse from a daytime Water-bird ceremony:

In the night time there is a big mob sitting round the fire – well they sing different than that, same *mura* but different way, you see. They call all the countries first, where the duck has been, where they lay eggs and one thing and another. And when they follow'm like that they sort of, they follow the country like in the history where the ducks have been travelling and when they get to the end like what we call *Thuntira, Thuntirayi nhanha*.

... Well that is the duck, then end of that part of the business. Well they go on again and call the country again. Different, different country, you know different names. Come from the same thing, see. Go like that, like in the night till it comes to this *mura,* finishing off. Well you finish off with that *mura* then where I sing it just now, *Palkakaranala Walkapityipityi* all that word, see. You know, the first one I sing. That is the finish. They had a lot of country to go over, name, name the country, take'm might be month before they can finish up, wait for the other people to come up finish the *mura* off. Whatever people can't come, of course they miss out.[10]

The idea of sharing the tradition with other groups was a most important feature of the *mura* stories. The word *mura* has a currency among a number of distinct language groups, in north-east South Australia and south-west Queensland, as well as well as the extreme north-west corner of New South Wales.[11]

Wangkumara people originally belonged to the area to the north of Tibooburra, and most of their country lay on the Queensland side along the Wilson River and part of the Cooper. Malyangapa people were further south including Cobham Lakes and over into South Australia, while Paakantyi people lived all along the Darling and the lower Paroo and through large areas particularly to the west of the river extending into South Australia to the west of Broken Hill. There were a number of distinct Paakantyi dialects, but basically it was one language and people from Bourke to Broken Hill and to Wentworth could all understand one another. The notion of *mura* however over-rode linguistic boundaries: it belonged to an area, rather than to a specific group. It was known to Wangkumara and Malyangapa people. Remarkably, it was known also to the northern Paakantyi speakers (Pantyikali and Wilyakali) who lived between Broken Hill, White Cliffs and Tibooburra, and to the Kurnu from around Bourke, but not to Paakantyi speakers along the Lower Darling River. Thus the word *mura* appears on its own as *muura*,[12] and in the phrase *muura maarni* 'Corroboree connected with Ancestors' in the Paakantyi dictionary,[13] but these words were only recorded from northern Paakantyi people, from speakers of Kurnu, Pantyikali and Paaruntyi. In fact, the northern Paakantyi, and particularly Pantyikali and Wilyakali, seem to have been influenced by the cultures of their neighbours to the north and west in a number of respects.[14]

In general terms, the *mura* type of story is characteristic of Central Australian cultures,[15] less so of the south-east, where there is less emphasis on travelling and where the narrative seems to be more dramatic.[16] In the Corner, as elsewhere, the *mura* may be snakes whose movements form the meander of creek beds and coils of sand hills; but they also include a bronze-wing pigeon, a kangaroo, euro, emu, dog and porcupine.[17] Their journeys may be confined to a locality or cover vast distances. Of the many stories we recorded we have selected the Two *Ngatyi*[18] ('rainbow serpents') for detailed presentation here.

Mura stories that have survived in areas of greater conservation further west often describe Ancestors travelling to a particular destination and then turning back to where they came from. Sometimes their journey has a distinct purpose:

Figure 6: Hannah Quayle outside her home in Wilcannia, 1957.
Photograph by Jeremy Beckett.

it is for collecting, by fair means or foul, some particular object. Thus the Lower Southern Arrernte Rain Ancestor *Anintyurla* travels far afield so that he can steal a particularly powerful rain-stone, and the Lower Southern Arrrernte Ancestor *Intara* goes to cadge from the reluctant Arabana owners a wonderful pounding stone, whose clear ringing sound he could hear from hundreds of kilometres away. This kind of *mura* story is known from the Corner Country too: George Dutton told Jeremy Beckett how people named *Drupada* went to Tibooburra to get a powerful rain-stone. Many Ancestors however just simply travel, creating and naming the country: the *Ngatyi* are typical of these. In the story as told by the two main story-tellers the *Ngatyi* go back to their own country in a subtle and indirect way: Birndiwalpi spring, where they finish, is linked to the Paroo where they came from, by underground channels, accessible only to the *Ngatyi*.

The *mura* text, as it has been communicated to us, may be no more than a string of place names, which can be recited, but the *mura* were originally performed at gatherings of people who came together to 'sing the country', as George Dutton put it.[19] People might not know all the names running the whole extent of a *mura* track; they performed the section they did know, giving way to those who could sing the next stretch, though by listening they might learn and eventually join in.[20] An important feature of these songs was that they passed through different language communities, but it does not seem that one was limited to singing one's own country. Dutton said that 'the song [that is the performance of the *mura*] might last for years and be carried on at other places' along the track.[21] We take this to mean that the people at one end of the track might not be present when the final phase was sung, although they might perhaps be entitled to participate.

There was an established repertoire of *mura:* these were passed from generation to generation. It seems that new *mura* songs called *maarni* could be composed, as a result of a dream: as Dutton explained, 'Your spirit travels. Then you make a corroboree about it. Others correct you on the names of places missed out. As you come to each man's country you have to act his *mura*.'[22] Hannah Quayle, who was an older contemporary of Dutton recalled how 'a Gungadidji[23] woman dreamed a song about the waterholes'. 'It came from a dead person and she smoked all her relations before she taught them'.[24] Perhaps most of these new *maarni* were forgotten after a while, but it may be that some were incorporated into the canon and the individual origin forgotten. From what Dutton said, the initial response to a *maarni* seems to have been to integrate it into the existing corpus of *mura*.

An ecological understanding of the *mura* stories might suggest that they functioned as maps enabling people to locate places vital for survival such as water holes. Alternatively, we can see them as animating the country, giving meaning to a landscape that, at least to European – though not we think to Aboriginal – eyes, seems almost featureless. The places one comes to, for one purpose or another, are not marked simply for their utilitarian value; they have stories that people know and can celebrate with their fellows. Moreover, the names which the *mura* gave to the places they visited on their travels are also the names of living human beings. At various stages in their lives both men and women were given the names of places on some *mura* track, and when they died these names could not be spoken for some years. Jeremy Beckett remembers Dutton weeping to hear the names of long dead friends when King Miller was singing the pituri *mura*.

1.2. General ceremonies gradually replacing the *mura*

In the Corner Country, towards the end of the nineteenth century, Aboriginal culture came under increasing siege by Europeans. People had to move to various centres – large stations, missions and townships often away from their own lands, and so the *mura* stories and ceremonies, with their detailed references to sites tended to be eclipsed by more general song and dance ceremonies, not closely linked to country. The *'Molonglo'* ceremony has been much discussed.[25] Mrs Moisey (Kurnu, that is Paakantyi from the Bourke area), speaking at an advanced age with Luise Hercus in 1972 recalled having seen it in the early years of the twentieth century.[26] There were a number of other general ceremonies, less well documented, which reached the Corner country. The *Nguninta* ceremony was certainly known there.

The situation with the *Nguninta* was very similar to that with the *Mulu,* as the *'Molonglo'* ceremony was called in the Corner Country. There can be no doubt that it is connected with the Arrernte *Unintha*. This name became *Ngunintha* in languages which do not allow initial vowels: in Wangkumara/Punthamara the name is pronounced 'Nguninta'.

The *Unintha* was at first known as an Arrernte women's dance. Spencer was not overly impressed by the dance, which he thought monotonous because it consisted simply of the women swaying their bodies in unison.[27] The distinguished old Arrernte man, Walter Smith in reminiscing about this ceremony to Luise Hercus most certainly did not share Spencer's opinion[28] and the ceremony was hugely popular also in Arabana country.

This ceremony survived for a long time among Arrernte people with men gradually becoming the main participants. In their dictionary Henderson and Dobson show this more modern development when they define the *'Unenthe'* as 'a type of ceremony where men dance. Women and children can also attend'.[29]

The form of the corroboree that reached the upper Cooper, via Boulia, was elaborate and anything but monotonous, with both men and women participating.

In May 1985 Jack O'Lantern and Cecil Ebsworth gave a detailed account of the *Nguninta*. They remembered it as it was performed in 1930 at Innamincka.[30] Jack could still sing many of the verses and gave detailed descriptions of the dancing. The main actor was said to have had more and more elaborate head-gear each night and after three weeks, by the last night, it was at its tallest and he wore varying neck-decorations.[31]

On a later occasion, speaking at Orange in 1989 to Isabel McBryde and Luise Hercus, Cecil Ebsworth recalled how the *Nguninta* arrived on the Cooper, imported from Boulia. He was only about 11 years old when he saw the Innamincka ceremony, but Jack O'Lantern who was many years older than him had taken part in the dancing

> Baryulah Bob[32] he was the one in charge of that, he must have went over there with them to Boulia (to learn the ceremony).

> There was a lot of women dancing and men too, they must have had different parts.

> That *Wakuli Wakuli nharra*, (a special final verse) that is the finish for the night every night.

> It might go off for two or three nights but then they would start up again and he would have another square round his neck, all feathers and grass − I wish I had cameras and that, it was lovely. It was at Innamincka, halfway between the town and the station, where all the graves are, they had big camp there. They danced the *Nguninta* for three weeks. They had other ones too, but this was the main one; a lot of Wangkumara too, they shake a leg, big mob, 30 or 40 people down there, young fellows and old fellows.[33]

There were other ceremonies, held particularly at Nockatunga and Innamincka, which may or may not have belonged to this category of 'travelling' ceremonies. Jeremy Beckett in 1958 spoke with King Miller, a senior Punthamara man living at Tibooburra. His notes state:

> **Gudja** corroboree in Innamincka, 1934. All naked. Coaxed up from Nockatunga. Miller was there.[34]

It is highly likely that the name represents the word *kutya* 'feathers' used in Mithaka and other languages to the west of Wangkumara-Punthamara and that this was a corroboree where only feather decorations were worn, as in the *Warrthampa* performed among Mithaka people.

**Figure 7: Cecil Ebsworth, Orange, 1990.
Isabel McBryde is sitting in the background.
Photograph courtesy of Isabel McBryde.**

These 'travelling' ceremonies are yet a further indication of interaction and cultural links between groups over a long distance. For people born after the turn of the nineteenth century, these ceremonies, which were spectacular and had great popular appeal, came to eclipse the *mura* stories.

1.3. Survival of the *mura* stories

The imported 'travelling' ceremonies were obviously popular, but they were not taken seriously by older people who were closely attached to their own country and who had learnt the *mura* stories. In the eyes of these older people, ceremonies like the *Nguninta* were just light entertainment. The *mura* stories meant so much more: they were the most treasured literature. It is those stories that the older people, born well before 1900 and particularly George Dutton wanted to talk about. It is those stories they told to Jeremy Beckett.

George Dutton recorded several *mura* stories on tape, and recited one (just a string of names) which he called the 'dark people's poetry'.[35] However, most of what we have was dictated in the form of a picaresque narrative, which was structured by the sequence of places along a track. This privileges the text and the mapping that is implicit in the place names, omitting the dimension of performance. It also privileges the individual *mura*-teller, whereas singing the *mura* had been an activity in which men and women shared their knowledge, and perhaps reached a consensus where recollections differed.

We have selected the Two *Ngatyi* or 'rainbow serpent' *mura* to present here, because it covers such a vast distance (from the Paroo River to the Flinders' Ranges, returning to a midway spot near the present day township of Tibooburra or in one version back to the Paroo); but also because we have three versions of it, and a fourth very brief reference which nevertheless includes an important detail. These versions, collected by Beckett in the winter of 1957, differ to some degree as regards the precise route the track follows, though they agree approximately as to where it begins and how far it goes, and agree precisely as to where it ends. We should however note the differences in the particular emphases and intentions of the *mura*-tellers.

Had the culture of which these *mura* stories were a part been still active, the telling of them might have been subject to a degree of standardisation, and it may be that no one would have been allowed to tell them on his or her own; however ceremonial life had fallen away in the early years of the twentieth century, and the senior men complained that the younger generation had no interest in such things. Remembering the *mura* stories in all their fine detail was no mean feat of memory under any circumstances; by 1957 it was a matter of reaching back into the past 50 or more years before.

1.4. The storytellers and their country

The *Ngatyi* traverse country belonging to two groups of Paakantyi people: the Paaruntyi from the Paroo and Pantyikali people, also called 'Wanyiwalku', further west. *Paaru-ntyi* literally means 'belonging to the Paroo', and refers to both the language and the people. *Pantyi-kali* literally means 'the Creek people'. *Wanyi-walku* is less easy to interpret: *palku* means 'speech', but what *wanyi* refers to is not clear. As often happens the language name is interchangeable with the name of the people. Pantyikali was a people's name that could also be used as a language name, while Wanyiwalku was a language name that could also be used as people-name. The two were not quite synonymous: in dictating a 'tribal distribution' for the north-west to Jeremy Beckett in 1957 George Dutton assigned a more easterly territory to the Wanyiwalku and a more westerly area to the Pantyikali:

From Yancannia Creek, Pantyikali, up to Yantara Lake and across to Mordern, to Wonaminta, Milparinka where they mixed with Malyangapa, then back to White Cliffs, where they mix with Wanyipalku. These go away to Momba and Piree Lake[36]

The languages that these people spoke were all Paakantyi languages, as mentioned above. They were mutually intelligible and really dialects of one language. Further on in their journey the *Ngatyi* traverse country belonging to people who spoke 'Yarli' languages, Malyangapa, Wadigali, and Yardliyawara, closer to the Flinders Ranges. These languages too were mutually intelligible and really just dialects of one language, but they differed considerably from the Paakantyi languages. The two main narrators were familiar with both these language groups: George Dutton was fluent in Paakantyi (Pantyikali) and Malyangapa, and Alf Barlow was fluent in the same two languages. Both had an intimate knowledge of large stretches of country on the New South Wales side of the story: George Dutton was particularly well acquainted with the Yancannia area, and this is where we have most detail.

Two of the storytellers, George Dutton and Walter Newton have already been the subject of articles by Beckett, and Hercus has also written about Dutton in various connections.[37] Hannah Quayle appears in the books of Bobby Hardy.[38] The other storyteller, Alf Barlow, has been mentioned in a general article.[39] All were aged between 70 and 80 or more when they were interviewed. All belonged to the back country between Broken Hill and Tibooburra, Hannah Quayle being born on Morden Pastoral Station, Newton on Tarella, Dutton on Yancannia, and Barlow probably also on Yancannia. Hannah Quayle regarded her main language as Malyangapa, though she also spoke the Paakantyi dialect Paaṟuntyi. Dutton when speaking with Beckett and with Hercus identified as Pantyikali, referring to the group of Paakantyi people from the creek country around Mootwingee [Mutawintji]. When speaking with Tindale in 1938 he only used the name Wanyiwalku ('Wainjiwalku' in Tindale's spelling) which was also used by Barlow and Newton. All the storytellers had been known to one another, but their life courses had diverged.

As a young man Dutton travelled around south-western Queensland and over to the Flinders Ranges and beyond, working as a drover but participating in 'corroborees' wherever he went. He could speak a number of languages, at least six fluently. Dutton prided himself about the depth and geographical reach of his knowledge, which he communicated without elaboration.

Unlike George Dutton, Barlow seems not to have travelled much outside the Corner, spending his middle years around White Cliffs, before settling in Wilcannia. He knew more than Dutton would allow, but alternated the *mura* stories with folk tales of *Ngatyi* and 'hairy men' (*makutya*). Hannah Quayle's life had been similar to Barlow's. She only referred to the Two *Ngatyi* story in passing, talking mainly about various events from her early life. Newton had a rather different way of telling the stories, which were mainly the dramatic Paakantyi stories, relating to the deeds of the Crow and the Eaglehawk, with only excerpts from the *mura* tracks. Unlike either Dutton or Barlow, he had attempted to weave these stories into a continuous narrative, which sounded at times like an Aboriginal Old Testament, describing how 'God' ('aka the Gulwiru') periodically chastised 'the people' for wrongdoing, and ending with 'God's' ascension, and the coming of white people.[40] The story of the Two *Ngatyi* appears in this narrative as a creation myth. Although of an age, the narrators were not personally close at the time Beckett took down their stories so that he interviewed each of them on their own; asking them to comment on the other's version would have given offence.

Beckett's work with Alf Barlow was limited to two long sessions in July 1957 on the Wilcannia 'mission'. He recorded a number of *mura*, as well as stories about *Ngatyi* and *Makudja* 'hairy men'.[41] Barlow died in 1962. Beckett had two sessions with Walter Newton in Broken Hill, in the July of 1957, and visited him briefly in the following January. He died in 1961. The Dutton corpus is very much greater, partly no doubt because he lived until 1968, but also because his knowledge of Aboriginal lore was extraordinary, and he was eager to communicate it, to white or black. Beckett worked with him at intervals from July 1957 to September 1967; he died in November 1968. Luise Hercus also worked with him over a few years, there being an agreement, stipulated by Dutton himself, that he would do linguistic work with her, and tell Beckett about *mura* and ceremonies. Over the years he provided detailed accounts of ceremonies, including the *milya* and the *wilyaru*, told several dramatic Paakantyi stories about the Crow, and numerous *mura*. Beckett found George Dutton a delightful companion, but as an informant he was dogmatic and at times testy. He cut short requests for more information with a peremptory 'That's the story!'.

Neither of us knew that he had once spent a couple of days working with Norman Tindale. It was on the Brewarrina Mission in July 1938, when Tindale visited there on the Harvard-Adelaide Anthropological Expedition. It was during a very low ebb in George Dutton's life, when he and his family and most other Aboriginal people had been forcibly moved from Tibooburra to the Brewarrina Mission, carted away in the back of trucks. Dutton loved Tibooburra: it was part of his 'Corner Country' – 'good place, Tibooburra' we have heard him say many times. The promised housing, for which George had already paid some money, did not eventuate. The mission was a long way out of town, among people from quite different groups. It was cold and wet, not like the Corner Country. It was like a prison from which George and his family managed to escape, not long after Tindale's visit, taking back-roads to evade the police.

Despite the dreadful conditions, George Dutton made a great impression on Tindale who wrote in his journal:

> G. Dutton of the Wainjiwalku tribe was my chief informant at this station. I have recorded more complete stories, an outline of the grammar and a series of comparative short stories in three associated western NSW languages to show the differences in style and vocabulary. These are all recorded in separate notebooks. Of one of his songs I obtained a poor wax reproduction on our wretched Edison recorder which gets worse every time we use it. This song is of special interest.[42]

This recording of George Dutton was the only recording made on this entire expedition.

1.5. The *Ngatyi* in the mythology of the Corner Country

Ngatyi is the Paakantyi name, the Malyangapa word is *kakurra*, which corresponds to *akurra* used in Adnyamathanha, the language of the Flinders Ranges, for what Australian anthropology has called the 'rainbow serpent'.[43] Although the particular manifestations vary a good deal across the continent, these supernatural beings seem always to be associated with water, whether as rain, and attendant thunder and lightning, or located in a waterhole, an underground spring, or in water flowing down a water course – whether an ephemeral creek or a river such as the Darling. As a generalisation it can be said that in Aboriginal belief a rainbow serpent is present wherever water is an active force in human existence, especially it seems where the source of the water is subject to seemingly mysterious variation. The *Birndiwalpi* Spring, which constitutes the axis of the story that follows, is stated to have an underground connection with the Paroo, many kilometres away.[44] Although water is beneficial to humankind, the rainbow serpent is regarded as dangerous if approached without the appropriate ritual precautions, and in some accounts can rear up, swallow and regurgitate intruders. Only a Clever Man can dive in a waterhole to bring on or stop the rains.[45]

Some, like the *mura*, perform their creative deeds and are transformed into features of the landscape: this is what happens to the Two *Ngatyi* in Hannah Quayle's version. However N*gatyi* may also be present in the here and now, and may be encountered by ordinary mortals. Alf Barlow said,

When the *Ngatyi* travel, they bring the water with them. When you see the water rushing up and coming a banker and shaking, that's the *Ngatyi* coming. When you feel water hot, *Ngatyi* been there. They only come after strangers: you've got to say, 'ngabeda angagu giredja'[46] – it's me, I've been reared here.

The paddle steamer (on the Darling) ran into a *Ngatyi*, one time. The *Ngatyi* makes a noise like a *bindi* (thunder) but not the same. At Bullea Lake, a few kilometres south of Mt Browne, a *Ngatyi* caught a man and swallowed him, stood up in the lake to swallow him, and then spewed him out. [Interestingly enough this lake is on, or very close to, the track of the *Ngatyi* as described by Dutton and Barlow.] At *Birndiwalpi* a boy went down and found it dry. Found a place like a crawfish hole, pushed a stick in and water gushed out. He stood there and found a snake coiled around his leg. He came up. His mother shouted, 'That's enough!' when it filled the flow stopped.[47]

In the myths of the Corner Country and usually elsewhere, these rainbow serpents are not included within the normal general term for 'snake'. They are always something special and different from ordinary snakes. For many people the rainbow serpents are still a source of fear and wonder. George McDermott, a slightly younger contemporary of George Dutton from the neighbouring Wangkumara people explained:

At some places, they reckon the *Parditha* [that is the *Ngatyi* in the Wangkumara language] is still sitting in the creek.

Where Jack Conrick has a place [that is at Nappa Merrie] and that is where that thing is. There is a big hole there, which never goes dry, *walya* (never). While they are alive them *Parditha* they never dry out. They got their own soakage and everything.[48]

The rainbow serpent figures in Aboriginal belief in two modes, it may be the tutelary spirit of a particular waterhole or a creator figure who moves over the landscape. Though the roles may be combined momentarily: in the mythology of the Corner Country in two separate stories, the *Guluwiru* (according to some a giant serpent) and the Bronzewing Pigeon (*mardi*)[49] both get water from waterholes putting the resident *Ngatyi* along with the water into their waterbags, releasing them further upstream, to run back to where they came from, or installing them in new waterholes. In the Two *Ngatyi* stories, however, while the *muras*' association with particular sources of water is important at the beginning and the end of the story, they are also travellers who form and name the country as they go. It seems that they cease their travels but it is not clear whether they or just their children become tutelary guardians of the water, either at *Birndiwalpi* or on the Paroo, or most likely linking the two.

Although the *mura*-teller may not have visited all the places mentioned in the narrative, the places are not imaginary. Beckett was unable to travel with his informants, but he was able to locate many of the places mentioned on a map, in discussion with Dutton. Now, by comparing maps checking Anglicised Aboriginal place names, and using linguistic data Luise Hercus has been able to establish not only the general direction of the journey, but also to identify more of the places mentioned in the story.

All four versions of the Two *Ngatyi* story agree as to the general direction of their journey, from east to west and back again, though in Hannah Quayle's short version, they do not return. Newton differs from Barlow and Dutton in having them return to the Paroo, but this variation is probably not significant, since one of the story's messages is to establish the subterranean link between the Paroo and *Birndiwalpi*, which was a spring north of Yancannia Creek.

Another important message of the story is the setting of some kind of a boundary at the edge of the Flinders Ranges. Both Dutton and Barlow agree on this, although one has people saying, 'Go back, this is not your country', and the other has the local *Ngatyi* saying it. Again this difference does not seem significant: the people and the *Ngatyi* of the place are equated. Hannah Quayle's version attributes the limit to the journey to the failure of water; it is hard to relate this to the other versions. All are agreed as to the approximate location of the westward boundary, but one might wonder 'why there?'. The location is not near any ritual boundary: the higher *wilyaru* rite extended from the Flinders Ranges to the Corner Country (though not as far as the Paroo). The boundary may have marked the westward limit of trading routes for the Corner peoples. According to Dutton, Aborigines travelled from the Corner Country over to Moolawatana to obtain red ochre, which they traded for possum skin rugs from the Paroo.[50] Moolawatana is not far from the place where the Two *Ngatyi* were turned back. The 'boundary' is clear however if one approaches the Flinders Ranges coming from the east, as did the *Ngatyi*. The Ranges rise straight and sheer out of the plain: there are scarcely any foothills. Only a few kilometres from the Paralana Hot Springs Mt Painter rises to 790 metres out of the plain. The *Ngatyi* traversed the Grey Range on their journey through the Corner: they got to Mt Browne – but that is a mere 274 metres and it rises from an elevated tableland. The eastern edge of the Flinders is one of the clearest natural boundaries one can imagine. It is a spectacular sight: it is as if one approached a vast stone fortress. It is the country of the Adnyamathanha, 'the Stone people', and their language belongs to a different subgroup from that of the Pirlatapa and also that of the Malyangapa and Yardliyawara people whose country the

Ngatyi have been traversing. George Dutton's *Ngatyi* are sent back at the Hot Springs, at the edge of the Ranges, while Alf Barlow's *Ngatyi* are sent back just short of the Ranges.

There were other *mura* stories that linked Paakantyi country with the eastern edge of the Flinders Ranges. The information available comes from Adnyamathanha people, and there is only scant mention of places to the east of the Ranges. D Tunbridge has recorded the story of the 'Two Mates', two male companions, who travel from Olive Downs, north of Tibooburra to the Flinders Ranges. They continue right across the Ranges and ultimately turn into a pair of Water-snakes, a male and a female.[51] Whether there is any remote link between this story and the *Ngatyi* myth is uncertain.

The story of the Goanna and the Native Cat goes from Poolamacca across to the Flinders Ranges at Prism Hill, south of Paralana. There is a large sandhill on one side of Prism Hill: that is where the Goanna dug deep down to try to escape from a vengeance party. J McEntee learnt about this myth from the Adnyamathanha elder, John McKenzie.[52] Prism Hill is known to Adnyamathanha people as *Vardna-Wartathinha* 'Goanna Deep down'. The story was noted by Mountford and published by D Tunbridge.[53] As in the case of the story of the 'Two Mates', very little geographical detail has survived of the eastern part of the journey of the Goanna and the Native Cat, and this makes the *Ngatyi* tradition even more remarkable.[54]

Only Dutton and Barlow attempt a place-by-place, name-by-name account of the journey, and both accounts become sketchy as the *Ngatyi* cross the South Australian border, though both have a name for the place at which the *Ngatyi* are sent back. Barlow's detailed knowledge fails earlier than Dutton's, possibly because he had never made the journey himself; he worked mainly as a station hand. Dutton was a drover and had taken stock over the border (on one trip he got as far as Finniss Springs). We do not know whether he had actually followed the *mura* track, but since the tracks usually typically link waterholes in arid country, this is quite probable. Across the border, and tracing the return trip, the two accounts are never far apart geographically. The greatest discrepancies are on the New South Wales side, with Barlow's route running well to the south of Dutton's, once the *Ngatyi* have left the Paroo, only converging, though still not overlapping, as they reach the border. The convergence is closer on the return journey back to *Birndiwalpi*.

It is difficult to explain the differences between two such detailed accounts. Dutton insisted that Barlow 'knew bloody nothing', but there seems to be no justification for regarding the latter's account as any less creditable. Nor in the face of so much detail, can we attribute particular differences to failing memory on the part of one or the other. It may be, however, that there had not for many years been a situation in which they had aired their rival versions. It may also be that their lives had been lived in different parts of the country, so that the story places they remembered from their travels were different. We like to think of the two detailed versions as different strands of the same tradition.

Endnotes

1. Beckett, Field notebooks 1957-8.

2. Gerritsen 1980: 11.

3. Compare with Hercus and Austin 2004.

4. As discussed by Breen 2007.

5. See for example Myers 1986.

6. Hardy 1969; Shaw 1987.

7. Shaw 1987.

8. See for example Beckett 1958, 1967. The Paakantyi practised two forms of male initiation called Dhalara and Dhamba. In the far Corner, as with the peoples further to the west, the Milia involved circumcision, Dutton and Barlow had both been circumcised according to this rite. Subincision was not practised in the region.

9. Horton 1996; Davis 1994.

10. George McDermott, recorded by Luise Hercus, Bourke Hospital, 29 October 1970, AIATSIS sound archive.

11. Elkin wrote of *mura-muras* linking the tribes from Birdsville and Innamincka, right down the Flinders Ranges and Port Augusta (Elkin 1938: 42). He elsewhere referred to the Aborigines of the New South Wales 'corner' as the easternmost extension of the Lakes Group (Elkin 1931: 51).

12. Paakantyi makes a distinction between long and short 'u' whereas the other languages using the term *mura* do not.

13. Hercus 1993.

14. Beckett 1958: 92.

15. See for example the work of Myers on the Pintupi and Dussart on the Warlpiri (Myers 1986; Dussart 2000).

16. These stories were mainly recorded by RH Mathews around the turn of the twentieth century, although Beckett recorded some stories in later years. For a re-analysis of these and other stories see Blows 1995.

17. A number of these stories are to be found in Beckett's note books. The bronze-wing story is published in Beckett 1979.

18. This is a Paakantyi word included in Hercus's dictionary. The equivalent word in the neighbouring Malyanghapa language is *kakurrra,* which corresponds to *akurra* in Adnyamathanha, the Flinders Ranges language, and *kakurra* in Nukunu from the southern Flinders Ranges.

19. Our informants did not say whether particular designs were associated with places, and we omitted to ask them.

20. Dussart describes a similar organisation among Warlpiri, with the songs divided into 'segments', and rights to sing them could be inherited or acquired (Dussart 2000: 20).

21. Beckett, Field notebooks 1957-8.

22. Beckett, Field notebooks 1957-8.

23. Kungardutyi was a name for Wangkumara people, immediately adjoining the Paakantyi and Malyangapa to the north, it originally meant 'circumcised' and was used by Wangkumara people to distinguish themselves from Karlali people who did not generally practice circumcision.

24. Beckett, Field notebooks 1957-8.

25. Confer with Mulvaney 1976: 90-92.

26. Annie Moisey, recorded by Luise Hercus, Wilcannia, December 1972, tapes 520-525, AIATSIS sound archive.

27. Mulvaney 1982: 40 and photograph.

28. Walter Smith pers. com. Walter Smith was 'The Man from Arltunga', see Kimber 1986.

29. Henderson and Dobson 1994.

30. George McDermott was initiated at that time – he was 'an old man' by then.

31. Jack O'Lantern and Cecil Ebsworth, recorded by Luise Hercus, speaking at Bennelong's Rest Home near Kempsey, 16-18 May 1985, AIATSIS sound archive.

32. This was Bob Parker Thinpika, a Yawarawarka man who spent most of his life on Nappa Merrie station. He was constantly in touch there with Wangkumara-Punthumara people, and they regarded him as one of their own. He was recorded speaking in Wangkumara 'Kungardutyi' by Bernhard Schebeck in c.1968.

33. Cecil Ebsworth speaking to Luise Hercus and Isabel McBryde, recorded by Luise Hercus, Orange, June 1989, AIATSIS sound archive.

34. Beckett, Field notebooks 1957-8.

35. Beckett, Field notebooks 1957-8.

36. Beckett, Field notebooks 1957.

37. Beckett 1958, 1967, 1978; Hercus 1982, 1993.

38. Hardy 1976, 1979.

39. Beckett 1994: 102-104.

40. See Beckett 1958, 1993.

41. Beckett 1992.

42. Tindale, Journal of the Harvard-Adelaide Expedition 1938-9: 177.

43. Radcliffe Brown 1986; McConnell 1930; Elkin 1930; Piddington 1930.

44. Shaw describes how Yancannia Station 'lay across the southern shore line … of the Great Artesian Basin' (Shaw 1987: 80).

45. Dutton pers. com.

46. This is a Paakantyi sentence, analysable as follows

ngabeda	nagagu	giredja
Ngapa itha	*yungkaku*	*kira-tya*
I here	own	country-having

    Here I am, it is my own country.

47. Beckett, Field notebooks 1957-8,

48. George McDermott, recorded by Luise Hercus, Bourke Hospital, 29 October 1970, AIATSIS sound archive.

49. Beckett 1978.

50. Beckett, Field notebooks 1957-8.

51. Tunbridge 1988: 86.

52. McEntee 1991 and pers. com.

53. Mountford, Journals and manuscripts relating to the Flinders Ranges, vol. 10; Tunbridge 1988: 41.

54. Particularly poor in detail is the story of the Seven Sisters (Tunbridge 1988: 16), adapted from Mountford. We are told that the events take place at 'Yandara in New South Wales'. This place is identified (p. 170) as 'Yandaroo Lake and Station, west of Bourke'. As there is no sizable lake on 'Yandaroo' presumably the reference is to Yantara Lake. No other placename is mentioned.

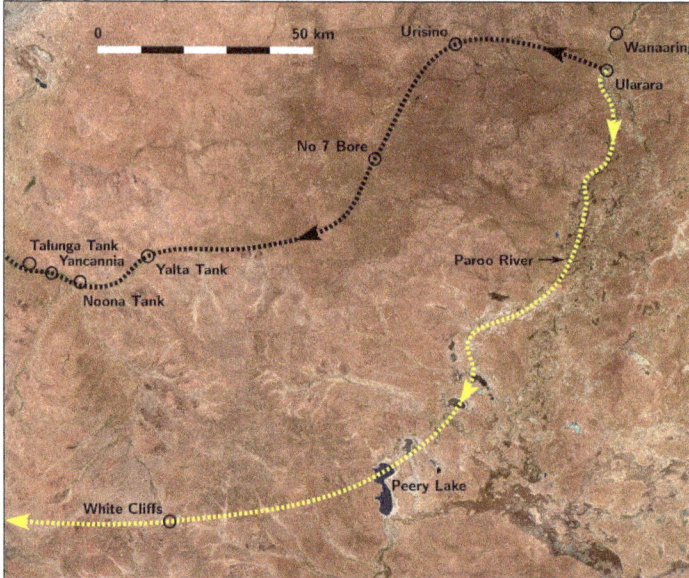

(black dotted line)	George Dutton's Ngatyi. Part of the route that is traversed once only.
(yellow dotted line)	Alf Barlow's Ngatyi. Part of the route that is traversed once only.

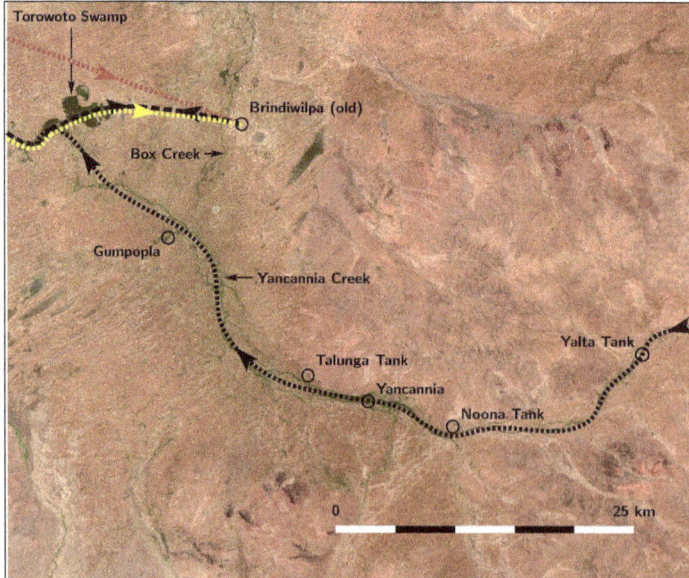

(black dotted line)	George Dutton's Ngatyi. Part of the route that is traversed once only.
(black dashed line)	George Dutton's Ngatyi. Part of the route that is traversed in both directions.
(yellow dotted line)	Alf Barlow's Ngatyi. Part of the route that is traversed once only.
(brown/red line)	Walter Newton's Ngatyi. Route traversed once only.

The two rainbow serpents: original texts 2

Texts (Sections 2.1. to 2.4.) are presented here exactly as recorded in notebooks by Jeremy Beckett in 1957-8. This means that the original spelling used in the notebooks has been retained here: it is based on the then newly practiced system of Arthur Capell.[1] In this spelling 'j' represents the sound of 'y' as in 'yes', and voiced consonants are used instead of the now preferred unvoiced consonants: that is 'g' is used instead of 'k', 'd' instead of 't' and 'b' instead of 'p'.

2.1. George Dutton's version

They started off from Ularada waterhole (that's on the Paroo 6-7 miles south of Wanaaring). They were Paroondji first.

Then they travelled across through Urisino station (before the station was there). The blacks must have noticed a lot of willows as marks to follow.

They went on past Number Seven Bore and they went into a place they called Yattla. (There's a tank there.) Here they talked Bandjigali. When they got there they was at the top end of the Creek at Yancannia station.

Then they followed the creek down. They came to a place they call Big Wunu and Little Wunu, after the little creeks come in about seven miles from the station.

Then they went down to a place they call Tartu wanyara [*tartu* is head].

Then they go down to Mingabulka (it means a dead hole as far as I can tell).

Well they went on a bit further then to a place they call Garrabala (there was a lot of sandstone there and they camped there one night and the old fellow called it Garrabala).

Then they went on to the next waterhole and that was Yancannia. They went on a bit further down to a waterhole they called Galbagarabuga (there's a quandong tree there, it was small when they came there). The old girl said, 'there's some *galbaga* (quandong) there!' So the old chap said, 'We call it Galbara buga.' Then they went down to a little waterhole in the bend of the creek. They saw some nardoo. She said, 'There's some nardoo'. 'Oh', said the old fellow, 'We'll call it Nardumuru.' (*muru* means black.)

Then they went on then and they saw a water hen trap, *malgaba*, just as if it had been made. They called the place Ngindatji (poor) malgabala.

Then they went on down to the next waterhole and saw a lot of *yarnda* (stones). then the old fellow turned to the old girl and said, 'We'll call this place Yanda:maga.' Then they went over to another place and saw a black oak growing in the water. Then the old woman said, 'There's a garlku there!' Then he said, 'Oh, then we'll call this place Galkwanbara.'

They went over and away to a fair sized waterhole, they called Ngaralba. They saw some sort of a bird, a little spotted one. They went over to another waterhole in amongst the stones. 'Oh, we'll call this Yarnandaya – a lot of stones.'

Away they went down the creek and they come to a place and had a bit of a feed there of moley apples. 'We'll call this place Maleanbala.'

Then they went on then to a place where they saw some caterpillars which might've bit them or raised a lump, they called it Yantabindaga (caterpillar nest).

Then they went on from there and came to a place they call Dalungga – it's a swamp.

They go on from there to where the *dinja* (Fog) used to camp. When he heard them coming he ran away. Then they said, 'Hallo, *dinja:bara.*' Then the old fellow called it Dinjawirga.

They went on from there to another waterhole and saw the willy-wagtail – *dindidindi*. The old fellow saw him up the tree and called the place Dindidindi ga murda (willy-wagtail's hole).

Then they went down then to another swamp and they seen a snake there. And he said, 'Oh we'll call it the snake brake, Duru gardu.'

Then they went on then to where the creek branched off and they parted there at Ngadji galbadidji (snake creek).

They joined up again at the place where they make the rain, Birndi walpi (thunder place).

Then they called another little place Banmara. I don't know what they saw there. Then another little waterhole they called Bilamora (pig weed).

The next place they came to they call Badu duru (*badu* is burr)

Then they went on from there and they saw a brake – like a snake place – they called it Gadu (brake) warlandi.

Then they went over and they seen a bird – like a bat – *diewaluru*, and they named that swamp Diewaluru.

They went on then, they seen a tree covered with green parrot – *gilguru*. Well they called that place Giburbula.

Then they went over from there to a swamp and they camped there for the night. Well the old man got up then, next morning, and saw her without a cover (it must have been summer time). He said, 'What made your old mick swell up?' They called that Munyu (vulva) baruda bagu (swelled up).

They went across to another swamp, joining that one and they called that Gaganganda (end).

Then they went across from there to Dilingga swamp – a big lake. The old woman got up and saw his two pricks. She said, '*dirli galbalu imamba*' (two pricks splitting laying).

Then they went over to a place a waterhole where they made a fire. 'We'll call it Wi:mani.' (*wi:* fire).

They went to another place and camped there. But the ants started biting them there so they called it Muningga (ants).

They came to Yantara Lake and saw two ducks. They called the lake Yulidinya (ducks-two).

Then they went on then to the end of the lake and came to some willows growing at the edge of the water. They pulled up there and had a feed. Them trees what they call *gamuru* and they called that place Gamuruganda.

Then they went from there to Yantara – *yantara* (stick).

Then they went on from there to a little waterhole they called Galili gurna (shit).

They went over from there across a salt lake [break] and to another lake they called Wonaginyi (boomerang broken) where a man had left a broken boomerang.

(Galti bula [two emus?] two lakes joined into one near Wanaginyi.)

Then they went on to another swamp further down, Meruna (*meri* – face) – could have been a rock or mud looked like a face.

Then they went on to another waterhole they called Bingara (some animal).

And they went on to another swamp and saw some reeds in the water. 'We'll call it Bulyubulyu.'

They went from there to Gurnana (shit) waterhole.

Then they went from there and saw a *gani* (lizard) up a tree. They called it Gani gutjana (lizard climbing).

They went up there to Mt Brown to a hill called Gambigambandi (blanket sewing) – they were Malyangapas and Wadigalis then – because they saw a man sewing a rug.

They went from there to Mt Arrowsmith which they called Binbili.

They went on and saw a *malbaru* (kite hawk) and called the place Malbuiu mida (kite hawk camp). [Marginal note: Quinyambie]

They went across to a lake called Gungumbi (dog).

Then they came to a waterhole called Mindilba, which brought them to Biriliba and Wadigali country.

Then they came to Dirabina (box tree big) swamp.

Then on to Balgiri (stream) waterhole.

Then on the main Yandama creek again and there they camped the night and went again in the morning. When they stopped the old girl said, 'Christ! We left our rugs behind.' So the old fellow said, 'We'll call it Gambi (rugs) gudianda (lost).'

Winadi lake. Ngamadangga – one tree, downstream from Yandama. Angoni. [Not clear where these all fit: Mt Poole – manibudadi. Didigarli. – (?bill's house). Dirawanda.]

When they get down to Bulka (beyond Yandama Lake) they travel down through a good bit of country but I don't know the names. Anyway they got about 50 miles on into the Walpi and Biralipa mob and some of the Dieri. They got on to the waterhole where there's a hot spring. At this point all the mobs there shouted, 'Go home to your own country.' Then they turned around and said

'Alright, *wilgamana*, but this hole got to be called Ngaba galala ganda (water hot).' That's why the water stayed hot.

[Note in margin: Galyamaru lake between Calabonna and Moolawatana in South Australia.]

They turned back from there. They came back practically the same places till they came to the place where they divided off – 'Ngadji galpadidji'. When you come to that places there's a little sandhill there they call Murlgulu (little bush). When they got there they stood up at each end of the hill. Just down from the sand hill is the waterhole Birndiwalpi. Their young they had left behind at Ularada waterhole had gone along underground and come up here. That's how the water got there. When they arrived there they saw their parents' shadows and started talking the Malyangapa language. They said '*Winea gulbila*' – 'Who's shadow's this?' '*Ingani nganu gulbiri.*' 'This is my shadow here.' Then they realised they were looking at their mother's and father's shadow. '*Yaga ngama bula dumabula*' – 'That's father and mother.' Then the parents said, 'What are you fellows doing here? – *Mina wandin dara. Ibani wirada* – go into the hole.' And they stopped there.

2.2. Alf Barlow's version

The she-*ngadji* laid an egg at Jularada waterhole before they set out. They set off along the Paroo, then they come to Piree Lake that's where they left the Paroo and made up towards White Cliffs way. They were crawling around near where the opal dump is. Those opals are the *Ngadji's* shit. Then they went out past the race-course and they were rolling about making a few swamps there and when they were going along they made a big wind that cleared the country and made it clear of trees. Then they got into a big gum creek and went past Noonthurungie Station and got up into a hill and made a big gap in the hill, but the old fellow said that this was not the place, and they went down the creek. At Blackfellers waterhole they saw a lot of blackfeller there, '*winbidja jabara*'. They came to Bancannia Lake and made it.

They went up then into the sandhill country. You can see their track $\frac{1}{m}$ about 30 yards wide. Then they made Starvation Lake and then over the sandhills again. (They must have known where they were going.) They rolled around in Lake Muck. Then they came to Bulka Lake. They went over the sandhills again.

They came to a waterhole and named it 'Dilarga'. They come to Cooney Bore and then across the sandhill again. They seen a big snake and when the snake seen them coming he ran away waving about and so made a sandhill 'Durugulili'. Then went again they come to a salt-lake, Maliga Lake. They went into the neck of the Salt Creek, 'Mangunguru'. Here some other Ngadjis stopped them and drove them back. They come straight back on the same road all the way back to Bulka Lake and went over a bit of a stony rise and camped on top of that hill formed a bit of a cane grass swamp on top of the rise. They were going down the Box(?) Creek and saw a *gani* (goanna) up a tree sunning himself and they named it 'Gani gudjanda'. They went on down the creek and the saw a wild dog, sitting down shitting and they called it Gunana. They went on and then they see snake track then – at Coally bore (on the Tibooburra road), and then they called that 'Durujaba'. Oh they said, 'he'll bite us we'll have to turn back'. Away they went down the flat country and then the old woman wanted to pee and the call it Dibara diriladja (pee emission). The same flat they came to a *widjadjugula* 'cocky'. 'Hello there's a cocky in the hollow tree'. They called it 'Gagalarinjabara ('cocky' hollow tree)'. They went down to a salt lake tank and they seen a bit of a spring there. There was some sort of animal there. They called that 'Warudarbaba'. They turned back then towards Yantara. They got down towards the neck of the Yantara Creek. They reckoned they saw a big blackfeller camp on one side. They made the creek deeper. They named the place 'Gumbagunbara (blackfeller)'. Then they went into Yantara Lake. They see a lot of duck then. 'Look at the *juli* there' – they called it 'Julidinja (duck lake)'. They went back across to where they came from and they come to a big waterhole and they called it 'Julunbugu'. They deepened the creek. They went on to 'Wimarni (fire burning)' Bore. Then to 'Dilinja' a box swamp. The old mother got a bit of a fright there and they called the lake 'Dingga'. They went down to another waterhole. They saw a shellparrot close to No 1. They saw these on a tree and called them 'Djirgowabulu'. They went down to Big Tree out in the middle of the creek and the old woman she got sick that night and swelled up in the belly. You can see the Box tree thick and round in the butt, standing up 'Murmurgali (sore behind)'. They went along again, Big Lake there. They call'm 'Ngunani'. They went along to Gagalbi and on to 'Muninga (ant's bed)', 'Diawaliru – they picked up a bit of blackfeller teeth'. They went down to the neck of the lake and saw an emu – 'Galidunggulu (stinking emu)'. They went down to Yantara swamp. Then to 'Murmurgali' (a different one). They went on and saw a *dinja* and called the place 'Dinjawirga'. Then on to 'Ngadji galbadidji (snake made a gutter here)' (where Number ?) goes away to Yancannia Creek. Then they went to Mungulbiri. They camped there one night and away they went. And then they could hear a noise. It was the young people singing

> bu:jebu:je banmarana
> bu:jebu:je baduduru
> jarawala namagulu
> gumagula dunggadunggarameida
> Juruljurul

They got down the spring and shut the young ones in the hole and got in themselves. The eggs had come down an underground hole in the Paroo. This was at Birndiwarlbi.

The old fellow go up a box tree way up. The young ones could see his shadow in the lake. He got off the tree and the old woman was waiting there. They fastened their tails together. Then they started into the lake in up the water-courses. He went the other way and then their heads met – they shifted whole sandhills while they were doing it. Then they pushed the young back into the spring.

2.3. Walter Newton's version

As noted by Jeremy Beckett, August 1957.

But while all the land was under water, God got the Holy Devil and his wife. This land was just flat with no drainage in it, no creeks, no rise or fall in the ground to make the creeks run either way; there were no rivers, no lakes. God gave the Holy *ngadji* instruction to bore out the lakes and to make a river. That's why the River Darling is like a snake's track – where they travelled along and bored a channel to make the river. They bored out lakes, they coiled around and scooped out the sand into sand hills. And they made channels to drain into the lakes, such as creeks. To make the water to run this way and that, they rose the ground up. Now after all this was done, right through Australian land, they lived at Peak Tank for a week, and that formed the Peak Hill. And they said, 'We'll go back to our children'. They went from there past Yancannia Station to a place called Birndiwalpa. When they came there they said, 'Oh we can hear the children playing.' And when then saw their parents coming they were excited and they ran down into the burrows.

And there was an ordinary big snake – poisonous – just beside the burrows – *muna* they call him, He's supposed to have been laying on his side singing songs – a *mura*. The two Holy *Ngadji* said, 'We'll sneak up on this *mura* and kill him.' However, something woke him up just as they were on him, and made down his hole. They just grabbed a couple of feet of his tail. That's how they got the [*dulpiri*] *mura*, and they dance a corroboree to that. They sharpened up emu leg bones and anyone who'd done wrong had to push that bone through his balls. This was because they'd eaten carpet snake before it was given to them to eat.

That hole was only the outdoor of the real house. They went down right through like a big rabbit burrow or tunnel and followed their children back to the Paroo. Everywhere was smaller then, God expanded the ground and creeks since.

2.4. Comment by Hannah Quayle

As noted by Jeremy Beckett, August 1957.

Jularada waterhole

Two *ngadji* travelled to Guliamaru in South Australia. The water was drying off. They died at the lake.

Barlow said *'galia maru'* meant 'bottomless'.

2.5. Comments by Cecil Ebsworth

The neighbouring Wangkumara/Kungardutyi people also had some knowledge of a *Ngatyi* story similar to the Dutton-Barlow-Newton one. This is evident for instance from the words of Cecil 'Knocker' Ebsworth (Wangkumara).[2] References to *Ngatyi* country are in bold:

> They had a song for all of them (the Water-snakes), big *mura*, see. He was a snake, they thought he was. He travel right through, right through Cobham Lakes, Salisbury Lake **all them Yancannia**, the *paritha* (the Wangkumara name for the *Ngatyi*).

George McDermott (Wangkumara) in 1971 sang a series of verses connected with the Water-snakes, but unlike George Dutton he did not tell of any distinct line of travel. The main Wangkumara tradition of the *'Paritha'* was clearly connected to country on the Cooper and probably had links to the *Ngatyi* story, but nobody recalled a continuous route in the way that George Dutton and Alf Barlow did. The Wangkumara tradition emphasises the idea that the line of travel of the Snakes is by subterranean channels, which are marked by the *kamuru* ('willow'-trees), George Dutton's text also mentions these trees as occurring near Yantara. Cecil Ebsworth continued:

> You can still follow the trees, though, one line of trees. Might be a couple of miles long, then might be only half a mile. If you keep going in that direction you might go ten or fifteen miles then you might see another one, going the same way. That is where the *mura* went.

> He start off right there near *Wiwilbura* waterhole, that is out from Ngakangura, (Nokanora waterhole on the Cooper) *Wiwilbura*, 'small wood'. He went from there, that is where the *kamuru* trees start from. He went right across then – funny thing – you go right across between Tibooburra and Yantara and you can see them right across go down Yancannia way, same lot of trees, same kind of tree. They never seem to

die. You can follow them right in the Cooper. You might see them only every two or three miles, but they are in the same line.[3]

Endnotes

1. Capell 1956.

2. Cecil Ebsworth senior was born in about 1919 at *Ngaka-ngura* 'Water-camp', that is the Nokanora waterhole on the Cooper up from Nappa Merrie. His parents named all their children after the places where they were born: the family travelled around a lot mainly on Durham Downs and adjoining stations as the father Albert Ebsworth worked as a station-hand and drover. 'Knocker' was simply an Anglicised version of *ngaka* 'water'.

3. Cecil Ebsworth, recorded by Luise Hercus, May 1985, Kempsey, AIATSIS sound archive.

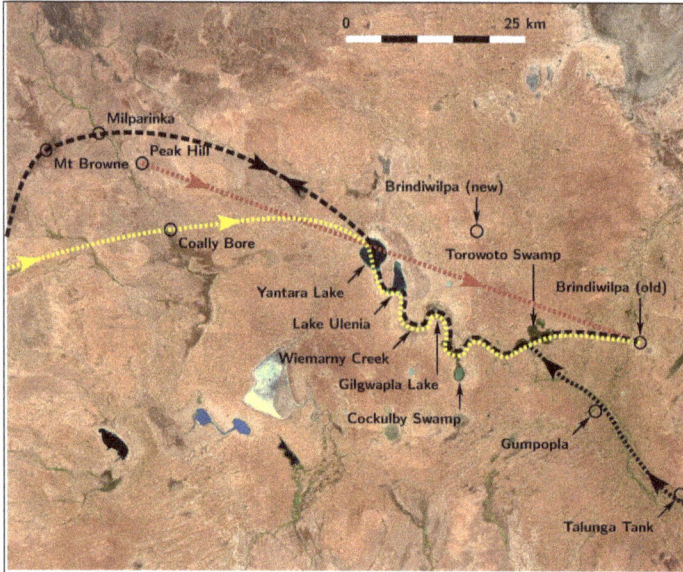

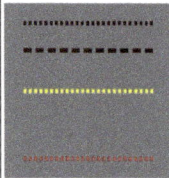

::::::::::::::::::::::::::::	George Dutton's Ngatyi. Part of the route that is traversed once only.
- - - - - - - - - -	George Dutton's Ngatyi. Part of the route that is traversed in both directions.
:::::::::::::::::::::::::::::	Alf Barlow's Ngatyi. Part of the route that is traversed once only.
::::::::::::::::::::::::::::::	Walter Newton's Ngatyi. Route traversed once only.

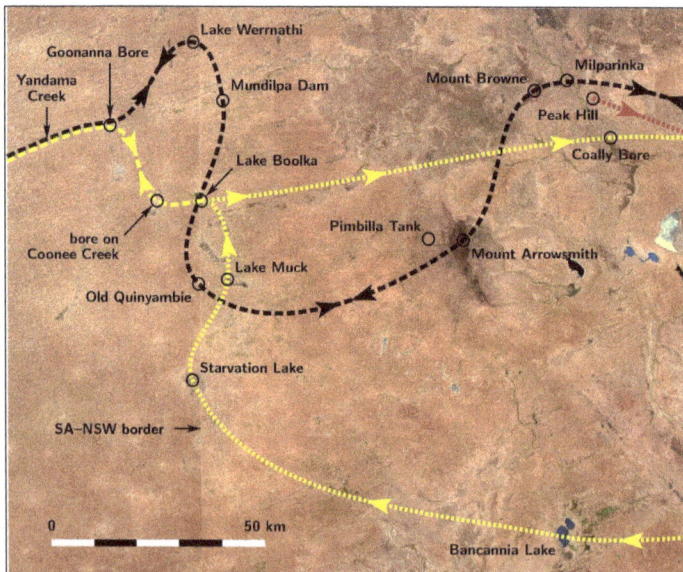

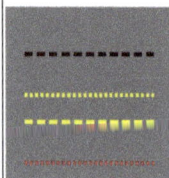

- - - - - - - - - -	George Dutton's Ngatyi. Part of the route that is traversed in both directions.
::::::::::::::::::::::::::::::	Alf Barlow's Ngatyi. Part of the route that is traversed once only.
- - - - - - - - - -	Alf Barlow's Ngatyi. Part of the route that is traversed in both directions.
::::::::::::::::::::::::::::::	Walter Newton's Ngatyi. Route traversed once only.

Geographical names in the Two *Ngatyi* stories 3

3.1. Introduction

There is a widespread popular opinion that Australian Aboriginal placenames all had some sort of descriptive meaning, and that names like 'Meeting of the Waters', and 'Meeting Place' were widely used. Much of the recent work on placenames, as for instance the paper by Peter Sutton shows that this was not the case.[1] There is great variation in the way placenames were formed, and there are regional differences too. One can generally find the following types of names:

1. Names that are just that: they cannot be analysed and their formation is lost in the mist of time. Attempting to analyse them is just like trying, in the absence of written historical records, to explain 'London' or 'Paris' in terms of modern English or French.

2. Names that refer to events in the travels of Ancestors. These are quite unpredictable, and when one does not know what the story was, there is no way one could attempt a correct interpretation.

3. Descriptive names: these may refer to earlier situations. There is no strict dividing line between this category and the preceding: an Ancestor may have seen a particular feature, which may or may not still be there, such as the moley-apple tree seen by the Two *Ngatyi* at *Maleanpala*.

Those names which are analysable can be quite short and consist of just one noun, as for instance the following which feature in the *Ngatyi* story: *Piri* 'Lake', i.e. Peery Lake, or *Malaka* 'String Bag', Lake Callabonna. They can be nouns with just a case suffix, as for instance the locative *–nga* in the name *'Muninga'* 'among the ants' from *muni* 'ant'. In some parts of Australia the use of locative suffixes as part of placenames is a dominant feature: it is well-known from the Adelaide area with placenames like Aldi<u>nga</u>[2] and has been studied in Northern Australia by Paul Black and Patrick McConvell.[3] Interestingly enough this use of the locative is found in only two of the *Ngatyi* placenames, *Muninga* and *Thirlinga*. It clearly is not a major factor in placename formation in the Corner Country, but it could have been an incipient development.

Particularly common among the names whose origin is attributed to the *Ngatyi* is the Paakantyi suffix *–anpala* 'with', as for instance in *Muuli-anpala* '(Place) with moley-apples'.

Compounds are frequently used, consisting of noun plus adjective, such as 'Nardu-muru' 'Nardoo-black', or two nouns, such as 'Dinya-wirga' 'the Fog's yamstick'.

The most striking names are those that consist of a whole clause from the story of the Ancestors who named the country, such as no. 25 'Ngadji-galbadidji', 'the Snakes parted company' and no. 73 'Gambi-gudiandi' '(our) rugs got lost'.

Unfortunately detailed knowledge of myths has faded over much of New South Wales and adjacent areas. One of the main contributions made by George Dutton and Alf Barlow in their telling of the myth of the Two *Ngatyi* is to illustrate the way in which Ancestors created and named the country. Traditional people had this very special vision of the landscape, and those two elders and the other contributors allow us to have a glimpse of that vision.

Both the main story-tellers had an intimate knowledge of large stretches of country on the New South Wales side: George Dutton was particularly well acquainted with the Yancannia area: this is where he gives us most detail, and information on many places that do not appear on any maps.

As discussed above (Section 1.4) the two main stories differ: George Dutton's *Ngatyi* travel west and return by the same route, but Alf Barlow's *Ngatyi* follow different outward and return paths on the New South Wales section of their journey. In oral traditions there is no right and no wrong version: narrators just follow different strands of the same tradition. Alf Barlow's version has had more adaptation to modern conditions: it includes the creation of opals. The two narrations have to be seen as just two versions of the same story, and as pointed out in Section 1.5 above, the ultimate points in the two accounts, the Hot Springs of George Dutton's story and 'the Neck of the Salt Creek' of Alf Barlow's story are not far apart.

Throughout one can feel the extreme intellectual honesty of the storytellers. When they don't know about the derivation of a name or are not sure they say so. Doing anything else would have been criminal in their eyes, it would have meant falsifying the past. They say simply 'I don't know what they saw there' or hedge their comments with phrases like 'as far as I can see', or 'maybe'.

3.2. Discussion of the placenames

Unless otherwise stated, the placenames given by the storytellers for the journey of the Two *Ngatyi* are given in the sequence in which they figure in the stories. For the main entry the original spelling has been preserved, but in the explanations we have used a modern practical orthography, which is given in italics.

Figure 8: Ularara Waterhole, 8 June 2008. Photograph by Sarah Martin.

Figure 9: Aerial view of Ularara Waterhole.
Photograph by David Nash.

1.   'Ularada' (Dutton), 'Yularada' (Barlow)

The transcription of Alf Barlow's pronunciation of the name is probably closer to the original, as Paakantyi does not have initial *u-*. The language spelling should probably be *Yulararda*. The name refers to the Ularara waterhole south of Wanaaring.

This waterhole was the place of origin of the *Ngatyi* in both the accounts, and therefore was a most important mythological site on the Paroo. It is likely that the name is the Paaruntyi form of a word attested for the neighbouring and closely related Paakantyi dialect Kurnu as 'yoolaroo ground of such wetness that the feet sink in it'.[4]

It seems that the area along the river south of the Ularara waterhole was also most strongly associated with the *Ngatyi*. The Nocaleche waterhole some ten kilometres to the south has its name from the *Ngatyi*: the placename Nocaleche represents

Nguku 'water' + *Ngatyi,* i.e., 'water belonging to the *Ngatyi*'.

This waterhole is now part of the Nocaleche National Park. The outward journey of the *Ngatyi,* as described by Alf Barlow, goes through this area, but the most important place, the original home of the *Ngatyi* was the Ularara waterhole.

Figure 10; On the Paroo; The Paroo Overflow at Momba Bore. Photograph by Simon Holdaway.

2.   'Peery Lake' (Barlow)

Peery Lake is one of the most important archaeological sites in New South Wales and has been studied by a number of archaeologists, including Isabel McBryde, and over recent years Simon Holdaway. It was THE LAKE par excellence for Paaruntyi people, hence it was called *Piri,* which is the Kurnu and presumably also Paaruntyi word for 'lake'. This word has been listed by Teulon for Kurnu:

Bee-ree   lagoon[5]

3.   'Big Gum Creek' (Barlow)

The 'Big Gum Creek' is probably the Bunker Creek, which comes down from near a gap in the rough ranges. Aboriginal people often spoke of creeks familiar to them by a description of the vegetation that grew along the banks, and instead of giving the actual name of a creek, they would describe it, saying for instance: 'You know that box-creek coming down?' These general descriptive terms sometimes came to replace an original specific name, and there are therefore in many parts of Australia toponyms like 'Box Creek', 'Gum Creek', 'Wattle Creek' and 'Gidgee Creek' in various spellings such as 'Gidgea Creek', 'Gidya Creek'. There is another example of 'Gum Creek' in the *Ngatyi* story in no. 60 below. It is likely that the modern name 'Bunker Creek' is linked in some way to 'Bunker', an Aboriginal man who died on Salisbury Downs in 1919 at the age of 90.[6]

The headwaters of the Bunker Creek are immediately adjacent to those of the Noontherangie Creek, and so this route fits in exactly with the 'creek-catching' travels of the *Ngatyi* (Section 1.5 above).

4.   'Nuntherangie' (Barlow)

This placename comes from another myth which George Dutton related in detail to Jeremy Beckett, that of the Euro and Kangaroo.[7] Those two Ancestors have a fight there over *ngarnti* tubers and over yams *nhanthuru.*

The name *Nanthuruntyi* is from the Paakantyi (Pantyikali) word *nhanthuru* 'yam', and the suffix *–ntyi* 'belonging to', i.e. 'belonging to yams'. This suffix is very commonly used to create proper nouns such as *Paaruntyi* 'belonging to the Paroo' and *Paakantyi* 'belonging to *paaka,* the river'. It is also used to form common nouns such as *yarrantyi* 'belonging to *yarra,* trees', i.e. a possum. The story of the fight over yams is associated with this immediate area.

Figure 11: Peery Lake. Photograph by Simon Holdaway.

Figure 12: Peery Lake. Photograph by Simon Holdaway.

**Figure 13: In the ranges near Bunker Creek.
Photograph by Simon Holdaway.**

5. 'Urisino Station' and 'No. 7 Bore' (Dutton)

There is no available information on the traditional names of these places. No. 7 Bore was originally a spring.

6. 'Yatla Tank' (Dutton), Yalta on modern maps

This is a site on Yancannia Ck. Old maps have the spelling **Yatala,** and that was probably the original name: George Dutton's pronunciation shows loss of the second −*a,* which was in an unaccented position. There is a possibility that this name is connected with the Paakantyi verb *yaata- yaata-la-* 'to search for something'.

The spelling on modern maps is 'Yalta', and there is also 'Yalda Downs' close by. These modern spellings probably represent a later and Europeanised pronunciation with metathesis, *lt / ld* replacing *tl.* George Dutton's evidence is important in this respect as it reinforces the earliest evidence that the original name was Yatala.

7. 'Big Wunu and Little Wunu' (Dutton)

George Dutton describes these places as being on Yancannia 'after the creeks come in seven miles from the station'. This is the exact location of **Noona Tank** on modern maps. There is therefore little doubt that the modern 'Noona' represents George Dutton's 'Wunu': the original name was probably *Ngunu*, *or Ngurnu*. The velar nasal consonant *ng* at the beginning of a word is unfamiliar to English speakers, and particularly when followed by *u* it is not clearly audible to English speakers. This would account for the differences in the European spellings. There are two waterholes in this location, one quite small.

Figure 14: Satellite photo of Noona Tank. From Google Earth.

The oldest available map reference, which is from Howitt, has the name as 'Nunno'. Howitt found the area attractive, and he wrote 'in spring it must be splendid'. George Dutton's description 'after the creeks come in' seems to refer to the same general area where the valley floor widens. Howitt describes this as 'the fall to Youngcanya'.

> On reaching the fall to Youngcanya the feed becomes very fine – plenty of grass and portulac everywhere.[8]

Howitt wrote this when he was camped at 'Nunno' waterhole before going on to 'Youngcanya', on his way to the Cooper to retrieve the bodies of Burke and Wills.

Note: _____

According to George Dutton's story the following three sites are all by the Yancannia Creek going down the very short distance of only a few miles from Noona Tank to Yancannia. They cannot be located precisely on maps. It is not surprising that they should be so close together: George Dutton knew this area particularly well, and it was a most fertile, and therefore no doubt a most frequented area on the route of the Two *Ngatyi*.

8.   'Tartu-wanyara' (Dutton)

'Tartu is head', said George Dutton. The Paakantyi word *thartu* 'head' also means 'hill'. It is therefore likely that this name refers to a small rise: there is one to the north of the Yancannia Creek between Noona Tank and Yancannia.

The meaning of 'wanyara' is not known, and there is a possibility that it was unanalysable, even to George Dutton.

9.   'Garrabala' (Dutton)

'There was a lot of sandstone there' according to George Dutton. There are several small European sandstone quarries on the eastern side of Yancannia station and it is likely that this site is in the location of one of them. The name can be analysed as Karra-(an)pala, '(Place) with sandstone'. The word *karra* 'sandstone' is not otherwise attested, but there is a well-known derivative: *karranya* means 'river sand' in all Paakantyi dialects, *–pala* is a short form of *–anpala* 'with'.

10. 'Mingabulka' (Dutton)

George Dutton said: 'it means a dead hole as far as I can see'. The context implies that this is a small waterhole between Noona Tank and Yancannia. The correct name was no doubt **Mingka-puka**, as *mingka* means 'hole' and *puka* means 'dead' in all Paakantyi dialects.

11. 'Yancannia' (Dutton)

This is a major waterhole and site of the old main station. The oldest spelling was 'Youngcanya' used by Howitt in his 1861 map,[9] Only slightly later are the spellings 'Uncana' and 'Uncanna', which are found on the old pastoral maps. There does not seem to be a convincing explanation for the name 'Yancannia', nor for other Paakantyi placenames ending in '-cannia'. The origin of the name **Bancannia**, which is on the route of Alf Barlow's *Ngatyi* is also unknown.

Note: ____

None of the following four sites can be located on modern maps, but from the sequence in George Dutton's story they are all waterholes in the Yancannia Ck going downstream from **Yancannia** to **Gumpopla**.

This stretch of the country was not quite as favoured as the area immediately around Yancannia, but even when the waterholes were dry there would have been soakages, and the sites are relatively closely spaced. 'Dalunga' (no. 21 below) should be on this stretch.

12. 'Galbagarabuga' (Dutton)

The name can be interpreted as ***Kalpaka-puka*** 'No good quandong', from *kalpaka* 'quandong' and *puka* 'dead', 'no good', 'rotten' (cf *Mingka-puka* no. 10 above).

13. 'Nardu-muru' (Dutton)

'There's some nardoo, we call it nardumuru. *Muru* means black' said George Dutton. This name is ***Ngardu-muru*** 'Black nardoo', from Paakantyi *ngardu* 'nardoo', *muru* 'black.

14. 'Ngindatji–Malgabala' (Dutton)

This name was explained by George Dutton as 'Ngindatji (poor) malgaba hentrap'. This name can be analysed as ***Ngiindatya-Malka-(an)pala*** '(Place) with a poor-looking waterbird-net'. The usual word for 'skinny, in poor condition' in all Paakantyi dialects is *ngiindatya*. The word *malka* means 'net' or 'duck-net', and *pala* is short for *–anpala* 'with' (see *Karra-(an)pala* no. 9 above).

In normal Paakantyi speech adjectives usually follow the noun, as in *Ngardu-muru* and *Kalpaka-puka* just above. If, however, one wants to put particular emphasis on an adjective, one puts it first; so someone who is really mad with a particular woman might say *'thurlaka nhuungku*, '(she is) a really bad woman' (*thurlaka* means 'bad'). ***Ngiindatya-Malka-(an)pala*** therefore means '(Place) with an absolutely pathetic-looking waterbird-net'.

15. 'Yanda-maga' (Dutton)

'They saw a lot of yanda (stones)'. This name can be analysed as ***Yarnda-maka*** 'Stone-hill', since *yarnda* is the word for 'stone' in the northern dialects of Paakantyi and *maka* is 'hill', cf 'mukko' Teulon.[10]

16. 'Galkwanbara' (Dutton)

'There is a galku (black oak, *belar*) growing in the water'. The name can be analysed as *Karlku-anpala* 'Black Oak-with'. This name has been further corrupted and appears as **'Gumpopla'** on modern maps, about 28 kilometres downstream from Yancannia.

Note: _____

The location of the next four sites is not known:

17. 'Ngaralba' (Dutton)

'They saw some sort of bird, a little spotted one'. This word is not known from other sources.

18. 'Yarnandaya' (Dutton)

They went over to another waterhole in amongst the stones. 'Oh, we'll call this *yarnandaya* – a lot of stones.'

This name is analysable as the reduplicated form of *yarnda* 'stone', **Yarnda-yarndaya** 'a lot of stones', with loss of a syllable in the seam of the compound, as in no. 24 below.

19. 'Maleanbala' (Dutton)

'They had a feed of moley apples'. The name represents Paakantyi **Muuli-anpala** '(Place) with moley-apples'.

20. 'Yantabindaga' (Dutton)

They saw some caterpillars, which might've bit them or raised a lump. They called it 'yantabindaga (caterpillar nest)'.

There is a possibility that the second part of this name represents the word written as 'pintooka', 'mat' by Teulon.[11]

21. 'Dalungga' (Dutton)

George Dutton described this place as a swamp. He gave no meaning for the name, so it was most likely not analysable even to him. It is now the site of the **Talunga** tank. The site is out of sequence, as it is well on the Yancannia side of Gumpopla,

Note: _____

George Dutton's Two *Ngatyi* are now approaching Birndiwalpi via the Torowoto Swamp. Birndiwalpi is their main site, it is their destination on the return journey. Alf Barlow's Snakes follow a similar track on their return journey.

22. 'Dinyawirka' (Dutton and Barlow)

George Dutton said: 'they go on from there to where the *Dinya* (Fog) used to camp … Then they said, "Hallo, *dinya:bara*." Then the old fellow called it Dinyawirga.'

'Dinya:abara' is Paakantyi **Dinya-*yapara*** and literally means 'dinya-camp', *yapara* being the Paakantyi word for 'camp'. There is loss of a syllable in the seam of the compound, as in no. 18 above: *Dinya-yapara> Dinya-para*.

'Dinya' is not the ordinary word for 'fog' in either Paakantyi or Malyangapa, but is the name of a mythological Fog Ancestor, who had his own story and was mentioned a number of times to Jeremy Beckett.

'*Wirka*' means 'yamstick' in northern Paakantyi (Teulon 'werrka')[12]: it seems that the Fog in his hurry left his yamstick behind, and so the place was called **Dinya-wirka** 'The Fog's Yamstick'.

Note: _____

The *Ngatyi* are now heading further northwest following the Yancannia Creek and they begin to talk in Malyangapa.

23. 'Dindindi ga murda' (Dutton)

'The old fellow called the place "Dindidindi ga murda" willy-wagtail's hole.' This name is analysable, but only in Malyangapa, as **Thindri-thindri-ka murda**.

Thindri-thindri is the Malyangapa word for 'willy-wagtail'. In Paakantyi the name of this bird is quite different: it is called *thirityiri*.

'Murda' is not attested elsewhere, and the function of the suffix *–ka* may be purely emphatic.

24. 'Durugardu' (Dutton)

They seen a snake there. And he said, 'Oh, we'll call it "the snake break", duru gardu'.

As mentioned in Section 1.5 above, it is important to note that the general term for 'snake', which is *thurru* in Paakantyi and Malyangapa, does NOT include Rainbow Serpents – they are different, and at no stage do the *Ngatyi* show any sense of affiliation with ordinary reptiles. Snakes hate strong winds, and so this is not an unusual name: there are several placenames in the Lake Eyre Basin that have a similar meaning.

Laurie Quayle, son of Hannah Quayle and brother in law of George Dutton, spoke to Luise Hercus about this site in 1974 in an unrecorded conversation. He gave the same derivation of the name as did George Dutton. He added that it was the *Ngatyi* who made the channels in that swamp, and one could still follow the way they went. The name, as he explained, is fully analysable in Malyangapa as **thurru** 'snake' + **kartu** 'windbreak'. In compound nouns the initial consonant of the second member of the compound is often lenited or lost (cf no. 18 above) and so the name was pronounced 'thurru-ardu'. This pronunciation is reflected in the spelling of the earliest references to this site by members of the Burke and Wills expedition. Beckler in his diary of 27 December 1860 spoke of the 'extensively branched marshlands which the natives call 'Duroadoo'.[13] Ludwig Becker drew a beautiful sketch, one of his last, of the group of Aboriginal people who came to see the exploring party at 'Duroadoo'.[14] The swamp was called 'Torowotto' by William Wright, another member of the expedition, who made the following sad comment about placenames:

> They (the local Aboriginal people) were well acquainted with the various creeks and named several placed[15] in advance, but our mutual ignorance of each other's language rendered it impossible to obtain any serviceable information.[16]

It is remarkable that in 1957, nearly 100 years later, George Dutton was able to give Jeremy Beckett the present placename-related information from this area, information that is not to be found on any modern maps.

The site is now called **Torowoto Swamp**. It became the location of a large sheep-station: it is well known to linguists because Curr's 1886 work contains a vocabulary from there by James A. Reid. Reid stated that the people there were called 'Milya-uppa' (i.e. Malyangapa), but the vocabulary he gives is entirely in Paakantyi. This is in keeping with what we learn from the Two *Ngatyi* who are speaking both Malyangapa and Paakantyi in the general area.

Note:

The route of the *Ngatyi* as described by George Dutton is not clear here, one would have expected them to go to Birndiwalpi first before going to Torowoto. It could be that George Dutton, talking about 'Dinya-Wirka', which was the Fog's camp, and then the Willie-wagtail's camp went on prematurely to talk about the Snake's camp.

25. 'Ngatji galbadidji' (Dutton and Barlow)

Then they went on then to where the creek branched off and they parted there at *ngatji galbadidji* (Snake Creek).

There are two possible explanations for this name, as there are two Paakantyi words *kalpa*, an intransitive verb meaning 'to split up', and a noun meaning 'small creek'.

1. The first explanation belongs to George Dutton's interpretation and is based on the Paakantyi verb *kalpa-* 'to split up': this same verb occurs in the explanation of the name **Dilingga**, see no.36 below. This Paakantyi verb can take the inceptive intransitive verbaliser *-rdi* 'to begin' and the Past Tense marker *−tyi*. The name is therefore analysable as:

    Ngatyi   kalpa-rdi-tyi

    Snake   split-start-PAST

    The Snakes parted company there.

2. The second explanation belongs to Alf Barlow's version and is connected to the noun *kalpa* 'creek'.

    'Ngatji galbadidji': 'snake made a gutter here, where Number ? goes away to Yancannia Creek.'

This name contains the Paakantyi word *kalpa* 'creek', the Paakantyi inceptive intransitive verbaliser *-rdi* 'to begin' and the Past Tense marker *−tyi*. The name is therefore analysable in Paakantyi as

Ngatyi   kalpa-rdi-tyi

Snake   creek-start-PAST

'The Snakes started forming a creekbed'.

It is quite clear that both George Dutton and Alf Barlow had a similar vision of the landscape: the Snakes separated, creating different branches of the creek. The two homophonous words *kalpa* in Paakantyi brought about the different explanations, one concentrating on the Snakes splitting up, and the other concentrating on the Snakes creating the branching of the creek. An old map

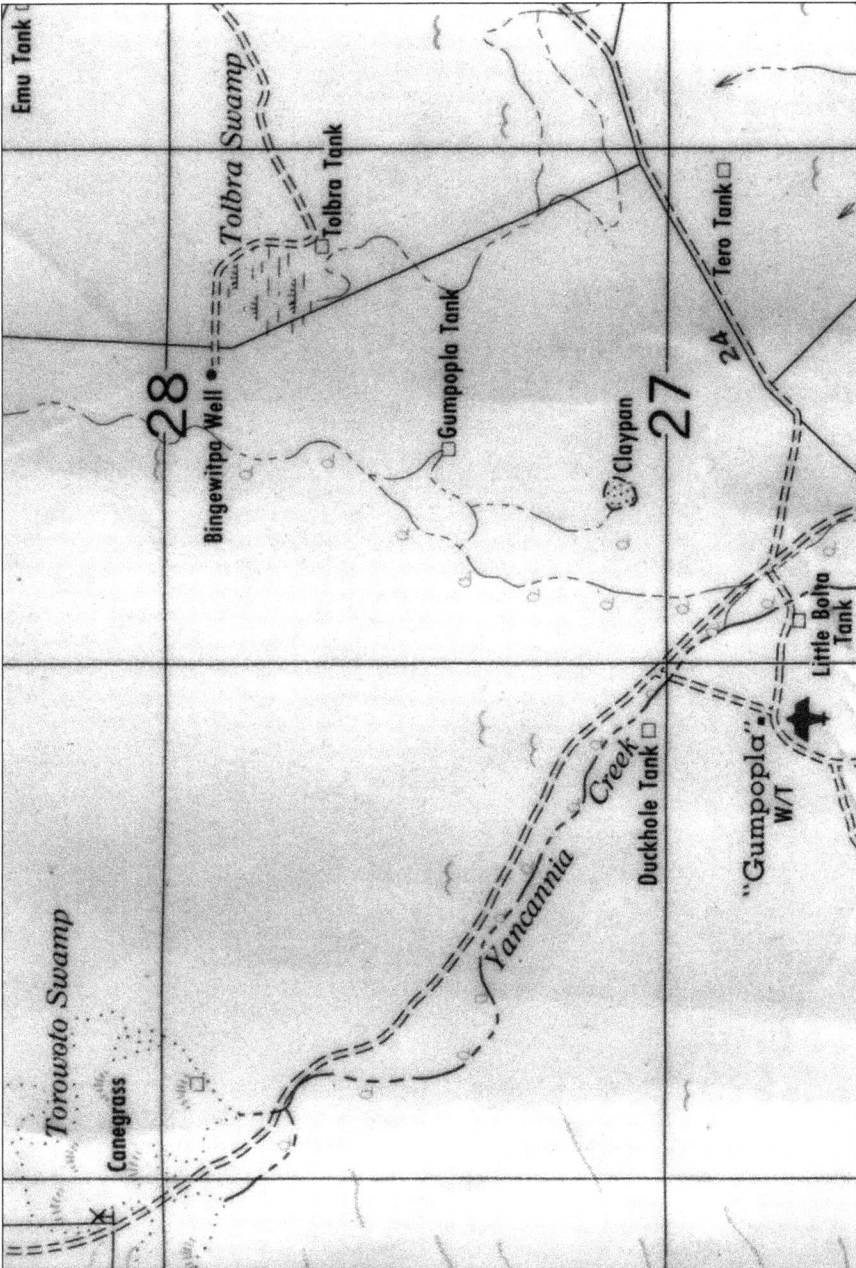

Text Map 1: Bingewitpa Well

Figure 15: On the edge of Lake Yantara. Photograph by Pamela Macdonald.

Figure 16: Canegrass Swamp in the Lake Yantara area. Photograph by Pamela Macdonald

'SH 54.12 White Cliffs' in the 1964 edition still shows 'Bingewitpa' Well. We think that the creek referred to is the one that passes the well on the western side.

26. 'Mungulpiri' (Barlow)

Barlow gives us no information on this site other than that it is between **Ngatji galbadidji** and **Birndiwalpi**.

'Mungulpiri' is a compound name, like most of the other placenames in the *Ngatyi* story, and it is analysable as Malyangapa *Mungu-kulpiri*.

As in the case of *kartu> artu* (no. 24 above) we have here an example of the loss of the initial g/k of the second member of the compound. *Kulpiri* is the Malyangapa word for 'shadow' and occurs in the final scene of the story in both the Dutton and Barlow versions, the scene that is played out at nearby Birndiwalpi (no. 83 below). The word *mungu* as such is not found in what we know of Malyangapa: there is however a verb *mungura-,* which means 'to be ignorant'. By mere guesswork we might suppose that the name means 'not knowing the shadow'. This would fit in with the story at Birndiwalpi (no. 83) where the young *Ngatyi* ask 'whose shadow is it?' Unfortunately we can't be sure of all this.

27. 'Birndiwalpi' (Dutton), 'Birndiwarlbi' (Barlow), 'Birndiwalpi' (Newton)

This is without doubt the most important site in the story of the Two *Ngatyi*: it is the place where there is a dramatic end. On the outward journey it is not so significant. We are told that the Two *Ngatyi* join up again at this site, having split up at *'Ngatyi galbadidji'* as is stated in George Dutton's account: 'They joined up again at the place where they make the rain, "birndi walpi" (thunder place)'.

The first part of the name could be analysed as both Paakantyi and Malyangapa: *pirnti* means 'lightning', 'thunderstorm' in both languages. The second part of the name, **'walpi'** has no obvious explanation in either language, in the present state of our knowledge. George Dutton translates 'Birndi-walpi' as 'thunder place', so there may have been a general term 'walpi' meaning 'place' or 'ground': the Paakantyi word *walpiri* 'bank, edge of a creek' may be a derivative of this noun. The spelling 'walpi' is well attested, as Jeremy Beckett noted it repeatedly from all the three main storytellers. This is in contrast to the European spellings, which have 'wilpa' and 'witpa':

The earliest spelling is from Tietkens **Pingiwilpi**.[17] Further variant spellings are: **Binjewilpa, Bingewilpa,** and **Bingewitpa Well** (on the 1964 White Cliffs mapsheet SH 54.12 as shown above, Text Map 1). Current maps no longer show this site.

The present day 'Brindiwilpa' station is near Yantara Lake and not in the location of the original site. This site was once a spring: 'a beautiful spot' according to Tietkens, who saw it in 1865,[18] but it was soon dug out and made into a well, and a bore was sunk nearby. The events taking place at this site on the return journey are described in no. 83 below.

Note:

> The location of the next five places is unknown: they must be between Birndiwalpi and the Yantara Lake area, presumably following the Round Lake Creek towards the west.

28. 'Banmara' (Dutton)

Panmara is a Paakantyi word meaning 'to scoop up water either with one's hand or with a dish'. 'I don't know what they saw there' said George Dutton, and so it remains quite uncertain whether this word for 'scooping up water' had anything to do with the name of the place.

29. 'Bilamora' (Dutton)

> Then another little water hole they call *bilamora*, 'pig weed'.

Pirla is a widespread word for 'pigweed' in Paakantyi, in Malyangapa and right over to the Flinders Ranges, cf.*virdla* in the Adnyamathanha language of the Flinders Ranges. In Adnyamathanha this plant is usually named *virdla-vaka*.[19] The second element, like the corresponding Parnkalla 'pakka', is a suppletive suffix that is added to some plant names. It is likely that in the name *Pirla-mura* the second part, *-mura,* is a suffix of the same kind, and probably does not specifically add to the meaning.

30. 'Badu duru' (Dutton)

> The next place they came to they call *badu duru* (*badu* is burr).

The word 'badu' is not known from other sources. As *thurru* means 'snake' in both Malyangapa and Paakantyi this name presumably can be interpreted as 'Snake-burrs'.

31. 'Gadu-warlandi' (Dutton)

They saw a break – like a snake place – they called it *gadu* (break) *warlandi*.

Kartu is 'windbreak', just as in *Thurru-kartu* 'Torowoto' no. 24 above. The rest of the name is unclear: *warlandi* means 'uncle', 'mother's brother' in Malyangapa. So could this be 'uncle's windbreak'?

32. 'Diewaluru' (Dutton), 'Diawaliru' (Barlow)

Then they went over and they seen a bird – like a bat – 'diewaluru' and they named that swamp 'diewaluru'.

This explanation presumably is based on a word for 'bat' that is otherwise unknown.

Barlow's interpretation is totally different: 'they picked up a bit of blackfeller teeth'. This explanation is based on the Malyangapa word *tiya* 'teeth'.

The situation here is the same as that for 'Ngatji Galbadidji' (no. 25), where the place and its name were firmly engraved in the minds of the narrators, but there were completely different explanations. These differences between the narrators show us quite how uncertain the 'meaning' of some placenames can be.

33. 'Gilgurbula' (Dutton)

'They seen a tree covered with green parrot – 'gilguru'. Well they called that place 'giburbula'.

This must represent *Kilkuru-(an)pala*, 'Green parrot -with'. The site appears as '**Gilgwapla Lake**' on modern maps.

'**Djirgowabulu**' (Barlow): 'They saw shell-parrot close to No. 1. They saw these on a tree and called them djirgowabulu'.

Both 'shell-parrot' and 'green parrot' are common names for 'budgerigar'. No. 1 refers to an old bore. Laurie Quayle pronounced the name of this bird *tyilkuru*.

In this case the explanations of the two narrators are practically the same, but the transcription differs just marginally: 'l' and 'r' are easily confused and so are gi/ki and dji/tji. Both narrators are referring to Gilgwapla Lake.

Note: ____

The next two places are also swamps, which are sometimes called ephemeral lakes. There are a number of these lakes in the general area south of Yantara Lake.

34. 'Munyu baruda bagu' (Dutton)

It seems from the story that this placename means *'munyu "vulva" baruda bagu "swelled up"'*. Nothing further is known.

35. 'Gaganganda' (Dutton) and Gagalbi Lake

They called that 'gaganganda' (the end, because it was at the end of the swamp).

The first part of this name is the Malyangapa word *kaka* 'head', *nganta* means 'to stop', 'to end'; so the name means 'head-end'.

The beginning of the name **'Cockalby Lake'**, immediately to the south of Gilgwapla Lake, contains the same word *kaka*, possibly followed by the word 'walpi' which is found in 'Birndiwalpi' (no. 27). There has been the usual process of elision at the beginning of the second word in the compound:

kaka-walpi > kaka'lpi

'head-place'

Cockalby Lake is mentioned as **'Gagalbi'**, without further comment, by Alf Barlow. There is a distinct possibility that 'Gaganganda', is an alternative name for 'Gagalbi', Cockalby Lake.

36. 'Dilinja' (Barlow), 'Dilingga' (Dutton)

This place clearly derives its name from the story: 'The old woman got up and saw his two pricks. She said, "dirli galbalu imamba" (two pricks splitting laying)'. The word *thirli* is 'penis'. *Imamba* is a form that is quite unmistakably Paakantyi: it contains *(ng)ima-* 'to lie down', and the Paakantyi second person singular intransitive subject marker *—mba*. In other words *ngimamba* is the normal Paakantyi way of saying 'you are lying down'. 'galbalu' is *kalpala* 'splitting' from the verb *kalpa-* 'to split' discussed under 'Ngatji galbadidji' no. 25 above. So what the old woman actually said was 'You are lying there with your split prick'. The actual form of the name should probably be **Thirlinga**, the final -*nga* is the locative marker.

Alf Barlow's explanation is presumably also from *thirli*, though Alf Barlow politely avoided saying so: 'Then to **Dilinja,** a box swamp. The old mother got a bit of a fright there'. It is characteristic of George Dutton's storytelling that he did not mince his words. The rude observation by the male *Ngatyi* at no. 34 parallels this comment from the female *Ngatyi:* it is all part of the story. As was the case in other areas, people would have laughed and enjoyed the story whenever such 'rude' placenames were mentioned - it was part of making the landscape come alive.

37. 'Wiimani', 'Wiemarny' (Dutton), 'Wimarni' (Barlow)

'A waterhole where they made a fire. We'll call it *wiimani'*, said George Dutton. Whereas the previous site had a Paakantyi name, this name is definitely Malyanapa. In Malyangapa **wii** means 'fire', 'firewood' and **mani**- means 'to get, to prepare', and so the name means 'making a fire'. George Dutton relates the name to a waterhole, but on modern maps the name refers to **Wiemarny** Creek, which enters Lake Ulenia from the south. The site probably is the main waterhole on this short creek, which links Gilgwapla Lake and Lake Ulenia. Alf Barlow translates the name as 'fire-burning' and identifies the site from a more modern feature: *'Wimarni* (Fire burning) Bore'.

38. 'Muninga' (Barlow), 'Muninngga' (Dutton)

'The ants started biting them there so they called it 'muningga', 'ants". The word *muni* means 'green ants' in both Malyangapa and Paakantyi, and *muninga* is the locative of this word in Malyangapa 'among the green ants'. This place is probably a site along the Wiemarny Creek closer to Ulenia Lake.

39. 'Galidunggulu' (Barlow)

Barlow explained that this name meant 'stinking emu'. The Malyangapa word for 'emu' is *kalithi,* and the word for 'stinking', 'rotten' is *thungka*. It is a clear example of loss of a syllable in the seam of a compound, as in no. 26 above. In this case the consonant *th* appears twice in succession, so one of the *th* sounds is lost by haplology, presumably the one at the beginning of the second member of the compound. The final vowel of the first word is also lost in composition:

kalithi+thungka > kalith(i)+(th)ungka

emu+stinking

The final *–(u)lu* is a suffix found in both Malyangapa and Paakantyi meaning originally 'one's own', 'one and only', 'just one'. So the name means ' just one rotten emu'. (In Malyangapa this suffix has become part of some kinship terms, for example *kakulu* 'brother').

Note: _____

Alf Barlow mentions four other places in the area, not referred to in the Dutton account. We do not know the exact location of these and the names are only partly analysable.

40-41. 'Murmurgali' (Barlow)

In Alf Barlow's account there are two sites in the Wiemarny Creek area, both named 'Murmurgali', which according to Barlow, means 'sore behind'. The chances are that this refers to a stretch of country with several soakages, two of which had trees with a swollen base: these represented the old woman *Ngatyi* with 'a sore behind'. The word 'murmu' does not appear in any of the available data, but the second part of the name, **kalhi-la**, is well known in Paakantyi as meaning 'to feel sore'.

42. 'Yulunbugu' (Barlow)

'A big waterhole, where they deepened the creek', said Alf Barlow. This probably still refers to the Wiemarny Creek.

There is some chance that this placename may mean 'long ago', *yurlupa* and *pukarru* are attested in Malyangapa and the closely related Yardliyawara respectively in the meaning of 'long ago'. The implication therefore probably is 'long ago (the *Ngatyi* deepened the creek)'.

43. 'Ngunani' (Barlow)

Barlow describes this as 'a big lake'. There is a Malyangapa word *ngurna-* which means 'to lie down', and *ngurnani* means 'you are lying down'. Unfortunately we cannot tell whether this word plays any part in the formation of the placename. The position of this 'big lake' is not clear, as there is no lake between the Wiemarny Creek and Lake Ulenia. It could be an area where the Wiemarny Creek widens out.

Note: _____

The two *Ngatyi* now travel across Lake Ulenia and Lake Yantara.

44. 'Yulidinya' (Dutton, Barlow)

George Dutton said: 'They called the lake *yulidinya* (ducks-two)'.

Yurli is the general term for 'duck' in both Malyangapa and northern Paakantyi. 'dinya' was not the normal dual marker in Malyangapa nor Paakantyi, but appears to have been an alternative, cognate with -*linya* 'pair' in Paakantyi.

Alf Barlow's explanation differs slightly: 'They see a lot of duck then. "Look at the juli there" – they called it 'julidinja', duck lake'.

There is no available evidence for 'dinja' meaning 'lake', and George Dutton's explanation seems the more plausible. The European version of this name '**Lake Ulenia**' is a good and clear representation of *yuli(d)inya* with the common elision of the initial consonant of the second member of a compound, as in 'Mungulpiri' (no. 26 above) and 'Galidunggulu' (no. 39 above).

45. 'Gamuruganda' (Dutton)

As George Dutton explained, the *Ngatyi* 'had a feed by some willow trees at the end of the lake'. These willow trees, gamuru / *kamuru* are the trees so closely associated with the underground channels of the *Ngatyi,* as described in Section 1.5.

This placename in modern orthography is **Kamuru-kantha** 'Willow-herbage', *kantha* is the word for 'herbage', 'grass' in Yardliyawara and therefore probably also in the very closely related Malyangapa. It is interesting to note that here, as at no. 19 'Maleanbala', where they eat moley-apples, the *Ngatyi* are depicted as herbivorous. Traditions (see Section 1.5) often describe them as carnivorous and even as man-eaters. The site is not marked on modern maps, but is presumably at the northern end of Lake Ulenia.

46. 'Yantara' (Dutton)

As George Dutton explained, **yantara** means 'stick' (in Malyangapa).

Note: _____

> The following seven places are not marked on modern maps, but from the sequence it is clear that they are in the vicinity of the northern end of Yantara Lake. Interestingly enough, six have Paakantyi names, and one is uncertain.

47. 'Galili Gurna' (Dutton)

George Dutton refers to this as 'a little water hole they call *galili gurna* (shit)'. This name contains the common Australian word **kuna** 'shit'. Although most of the names are Malyangapa at this stage in the journey, there is a likelihood that

'Galili' is based on the Paakantyi word for 'dog', **karli** followed by the suffix –*ili* 'now', 'here now', and that the name means 'there is dog-shit here now'. The place is not marked on modern maps.

48. 'Wonaginyi' (Dutton)

George Dutton mentions 'another lake they called "wonaginyi" (boomerang broken) where a man had left a broken boomerang'. **Wana** 'boomerang' is a widespread word and is found in both Paakantyi and Malyangapa. 'kinyi' is Paakantyi *(k)inhi / (k)inyi* 'this here'. The name therefore means 'there's a boomerang here'.

49. 'Gumbagunbara' (Barlow)

Alf Barlow explained:

> They got down towards the neck of the Yantara Creek. They reckoned they saw a big blackfeller camp on one side. They made the creek deeper. They named the place 'Gumbagunbara (blackfeller)'.

Kumpa-kumpara is a well-attested word in Paakantyi: it means 'a lot of (Aboriginal) women'. This is not contrary to Alf Barlow's interpretation: there would have been many occasions when only women were left in camp, and the term 'blackfeller' applies to both men and women.

50. 'Galti bula' (Dutton)

George Dutton described this as 'two lakes joined into one'. This must refer to the double lake, to the north of Yantara Lake and connected to it by a narrow channel, the 'neck' mentioned by Alf Barlow. The double Lake has no name on the map.

This is a Paakantyi name **Kalthi-pula** 'Two Emus', from *kalthi* 'emu', and *pula* 'two'.

Note:

> The following two places, not found on maps, are mentioned by Alf Barlow as being in the flat, flood-prone area between Coally and the west side of Yantara Lake. In Alf Barlow's account the two *Ngatyi* travel through this area on their return journey.

51. 'Dibara diriladja' (Barlow)

Alf Barlow interpreted this as 'pee-emission'. **Thipara** is the word for 'urine' in both Paakantyi and Malyangapa. **Thirila** 'flow' (noun) is not attested in Paakantyi, but is known from a song in the neighbouring Wangkumara/Kungardutyi language, a song to stop heavy rain, sung by Jack O'Lantern for Luise Hercus:[20]

Wapaya   thirila

stop   flow

'stop the flow'

The suffix −*tya* in Paakantyi means 'having', 'full of'', and so the placename means, exactly as Alf Barlow implied, 'having a flow of urine'.

52. 'Gagalarinjabara' (Barlow)

In Alf Barlow's account the two *Ngatyi* came to a 'widjadjugula', 'cocky':

'Hello there's a cocky in the hollow tree.' They called it Gagalarinjabara ('cocky' hollow tree).

'Widjadjugula' should be **'witya-tyukura'** as *witya* is the Malyangapa word for 'white cockatoo', and 'tyukuru/tyukura' is the Malyangapa word for 'hollow tree': this is confirmed by the Adnyamathanha (Flinders Ranges) word *(wira) ukuru* 'hollow tree'.[21] The meaning is therefore 'a white cocky's hollow tree' and the language of this statement by the *Ngatyi* is Malyangapa.

As the white cockatoo (corella) is by far the most common species, the term 'wiltya' is used here as a general term: the placename 'Gagalarinjabara' tells us that it is not a white cockatoo that lives in the hollow tree, but a Major Mitchell cockatoo, which is very rare in the area. The term *kakalarinya* is known from Teulon's vocabulary as 'Kahgoolarinya, tricolor-crested cockatoo (Leadbeater's)' i.e. Major Mitchell cockatoo.[22] The second part of the name is *yapara* 'camp' (see no. 22 above). This placename can therefore be interpreted as **Kakalarinya-(ya) para** 'Major Mitchell cockatoo's camp', with the usual elision of the beginning of the second member of the compound. This name is Paakantyi: the Malyangapa word for 'camp' is *ngatyara*. It therefore seems that at this spot the *Ngatyi* are talking about the cocky-tree in Malyangapa, but they are naming the site in Paakantyi.

Significantly, this is among the last Paakantyi names on the westward journey of the *Ngatyi*.

59

53. 'Merina' (Dutton)

George Dutton suggested that this 'could have been a rock or mud that looked like a face.' Both Paakantyi and Malyangapa have the word *miri* 'face', but the form *miri-na* 'his/her face' is Paakantyi.

54. 'Warudarbaba' (Barlow)

According to Alf Barlow this place was named after 'some sort of animal' that was there.

There must be a connection between this name and the following 'Bingara'. All we know of these two sites, one mentioned by Barlow and one by Dutton, is that they are on the eastern side of Coally Bore. The two 'animal' names could be referring to adjacent sites or even the same site known by two different names.

55. 'Bingara' (Dutton)

George Dutton, in almost identical terms to those used by Alf Barlow for Warudarbaba (above), explained that this name referred to 'some animal'. Nothing further is known. This is a striking example of the two storytellers telling the same story, but according to somewhat different traditions.

56. 'Bulyubulyu' (Dutton)

According to George Dutton the two *Ngatyi* here 'saw some reeds in the water'.

The word 'bulyu' has been recorded for Paakantyi by C Richards as 'bpool'yoo, bog'.[23] Richards also lists what is obviously a derivative *pulyuru/pulyura* 'bpool'yooroo, bpool'yoora swamp'. This is the widespread word *pulyurru* 'mud', found in languages further north and northwest, in Yandruwandha, Diyari and in Wangkumara. It seems therefore that the most likely meaning of the reduplicated form **'Bulyubulyu'** is 'a swampy place'. (For different effects of reduplication see no. 18 above and no. 62 below).

57. 'Durujaba' (Barlow) at Coally

> They see snake track then – at **Coally Bore** and then they called that 'durujaba'.

Coally Bore is by the Evelyn Creek and the 'track' no doubt refers to the winding channel of the creek. *Thurru-yapa* is the Paakantyi word for 'snake-track', for *thurru* 'snake' see Durugardu no. 24 above. *Thurru* is an ordinary snake, not a *Ngatyi*, and so it seems that the *Ngatyi* did not create this creek but that it was made by the track of an ordinary snake, and the *Ngatyi* came there and named it.

The Coally Bore is mentioned only incidentally here, as being by the *Thurru-yapa* creek, Evelyn Creek. Coally has a Paakantyi name, perhaps shared with Malyangapa, *Kuuwali,* which refers to a species of owl. This owl is apparently not connected with the *Ngatyi* story.

**Figure 17: Mintiwarda, Peak Hill.
Photograph by Pamela Macdonald.**

58. Peak Hill (Newton), Mintiwarda (Ebsworth)

This conspicuous site is not referred to in either George Dutton's or Alf Barlow's version of the Two *Ngatyi* story. George Dutton however mentioned the name *Mintiwarda*, 'Peak Hill' in his story of the Kangaroo and Euro.[24] The site is only about eight kilometres north-northwest of Coally Bore, and is right by the Evelyn Creek. It is the main site described by Walter Newton: the only other place Newton mentions is Birndiwalpi. He relates how the *Ngatyi* dug into the ground here to make channels, and piled up the soil to create hills: 'they lived at Peak Tank for a week, and that formed the Peak Hill.' Walter Newton had obviously heard a somewhat different version of the two *Ngatyi* story from that told by George Dutton and Alf Barlow, who do not mention Peak Hill. Cecil 'Knocker' Ebsworth (Wangkumara) however must have heard a version of the story similar to the one told by Walter Newton. He knew the name *Mintiwarda* and mentioned two snakes 'sitting about' in the area. The name *Mintiwarda* is not analysable except for the first part ***minti,*** which means 'nose' in Malyangapa.

Note: _____

The next two sites are not known from modern maps, but they must be in the area between Coally Bore and Mt Brown. They are part of George Dutton's description of the outward journey *and* Alf Barlow's description of the return journey of the *Ngatyi*.

59. 'Gurnana' (Dutton), 'Gunana' (Barlow)

The Two *Ngatyi* called this place the 'Shit' waterhole, because they saw a wild dog defecating just there. The widespread *kuna* 'shit' word is common in placenames: no. 48 Galili-gurna is identical in meaning.

The word **kuna** is used in both Paakantyi and Malyangapa: the name is more likely to be Malyangapa because the two *Ngatyi* have now switched to this language.

60. 'Gani Gudjanda' (Dutton), 'Gani Gutjana' (Barlow)

Both George Dutton and Alf Barlow had identical explanations for this placename: it means 'lizard climbing' in Malyangapa. Alf Barlow, referring to the return journey said:

'They were going down the Box Creek and saw a *gani* 'lizard' up a tree. They called it *Gani Gutjana* 'lizard climbing'.

Figure 18: View from Mt Poole. Photograph by Pamela Macdonald.

This creek is probably the 'Gum Creek' shown on maps as coming down from Mt Shannon. This name is analysable exactly as said by the two speakers. ***kani*** is the word for 'frill-neck lizard' in Malyangapa and a number of neighbouring languages. ***kutya-*** means 'to climb' in Malyangapa: *-arnda* is the present tense suffix, *-arna* is a participial suffix.

61. 'Manibudadi' (Dutton)

This name, from the way it is written in the notebook, refers to Mt Poole. This prominent hill is named later on, as a kind of afterthought in the account of the journey of the *Ngatyi* when they travel through an area near the South Australian border. Geographically the name fits into the sequence here, on the way to Mt Brown. The name is not clear: *mani-* means 'to get' in Malyangapa (see no. 37 above).[25]

62. 'Gambigambi' (Dutton)

In George Dutton's account this name, explained as 'Blanket sewing', was given by the two *Ngatyi* to a hill near Mt Brown, because they saw a man there, sewing a rug. The word *kampi* for 'rug' or 'clothes' is known over a wide region, including both Paakantyi and Malyangapa speakers. The placename is a reduplicated form of this word *kampi*. In all the languages of the general area reduplication of nouns can imply a diminutive or deprecating nuance, and so *kampi-kampi* literally means 'a little rug' or 'some sort of rug'.

Note:

It is here that George Dutton remarks on the languages that the *Ngatyi* now spoke: 'they were Malyangapas and Wadikalis then'.

63. 'Binbili' (Dutton)

In the story the two *Ngatyi* next went to Mt Arrowsmith, which they called 'Binbili'. Nothing further is known of this word.

Fortunately the name has not disappeared: it figures as 'Binbila' on modern maps, not as the name of Mt Arrowsmith itself, but as the name of a nearby tank by the side of a creek which presumably also had this name.

64. 'Malbulu-mida' (Dutton)

This name is interpreted by George Dutton as 'Kitehawk camp'. It appears to be a Pirlatapa name, or Wadikali strongly influenced by Pirlatapa. The Pirlatapa people and their language were called 'Biraliba' and alternatively 'Biriliba' by

George Dutton. The language was part of the close-knit Diyaric group, which included the immediate northern neighbours of the Pirlatapa, ie. the Diyari and the Yandruwandha.[26] This placename is clearly Diyaric.

The first part, given by George Dutton as 'malbaru, kitehawk' is known as a bird name *malparu* from the Strzelecki form of Yandruwandha, adjoining Pirlatapa. The translation given by Breen, from only one informant is 'flock-pigeon'.[27] AW Howitt, quoted by Breen, has yet another definition of *malparu*, as 'crane, black with white on wings'. Howitt's description doesn't fit in with George Dutton's 'kite-hawk' any more than does a flock-pigeon. Reuther in his placenames volume explains the name of another place called 'Malparu 'as 'a big bird':

> Here Watapajiri's men spotted a karawora (= 'eagle'), and exclaimed: "that's a big bird!" this was the reason for the naming (of Malparu).[28]

It therefore seems that in Diyari and Pirlatapa **malparu** was the name of a raptor.

Mountford in his collection of data, which includes information from east of the Flinders Ranges, has an entry '*wildu malparu* = white eagle' (*wildu* is the word for 'eagle' in Adnyamathanha, the Flinders Ranges language).[29] By far the most detailed information comes from John McEntee, whose station, Erudina, is not far from the route of the *Ngatyi,* and who has been able to observe birds in company with Adnyamathanha elders. He has *marlpuru wildu* 'Spotted Harrier, Circus assimilis'.[30] This vindicates George Dutton's explanation.

The second part of the name **mitha**, means 'ground', 'place' in Pirlatapa and Diyari. This placename therefore means 'the Kitehawk's place', i.e. as George Dutton put it, 'his camp'. There is a well-known parallel to this placename on the Birdsville track, Mira Mitta Bore, which is *Mayarru-Mitha* 'Rat place', 'Rat camp'.

65. 'Gungumbi' (Dutton)

'They went across to a lake called *gungumbi* (dog)', said George Dutton. This is certainly Quinyambie, which is written in the margin of J Beckett's notes at this place. **Quinyambie** is the name of a now deserted homestead just on the South Australian side of the border. The name is based on the Malyangapa-Wadikali word for 'dog', which was **kunyu**. The derivation of the second part of the name is not known.

Figure 18a: The Old Quinyambie Homestead. Photograph by John McEntee

Note: _____

> Just after this stage George Dutton states that the *Ngatyi* had come to 'Biriliba and Wadikali country' and this is borne out by the Pirlatapa placename no. 64 Malparu mita, which he had just mentioned. His skill and knowledge was uncanny.

66. 'Didigarli' 'Bill's House' (Dutton)

This name is not analysable: the first part could be the verb *thithi-* 'to see' in Malyangapa-Wadikali.

'Bill's House' which is mentioned along with 'Didigarli' is an Aboriginal version of the English name **'Smithville House'**, an old border post between South Australia and New South Wales just north of Boolka Lake.

67. 'Mindilba' (Dutton)

In George Dutton's account the Two *Ngatyi* came to a waterhole called 'Mindilba'. Nothing is known of the derivation of this name: it is likely to be a compound with the second member being *wilpa* 'opening' in Pirlatapa. From the evidence

Figure 19: Hearths revealed by erosion on an extensive Aboriginal place on Yandama Station recorded by Isabel McBryde in her 1963 survey. Its archaeological features included surface scatters of stone artefacts, hearths and stone arrangements (site YA 1). Ranging poles marked in feet. Photograph by Isabel McBryde, August 1963. Digitally enhanced by RE Barwick, 2007.

of the story it is clear that here the *Ngatyi* are heading north, and this is a waterhole in the **Mundilpa Creek** north of Smithville House, by the Hawker Gate. The Mundilpa Creek runs roughly parallel to the Yandama Creek and is only a couple of kilometres away from it to the south.

Note:

> From the account given by George Dutton it seems that the following three sites are also on the Mundilpa Creek, as the *Ngatyi* are heading further into South Australia. Unfortunately no exact locations are available.

68. 'Dirabina' (Dutton)

The name of this swamp. according to George Dutton, means 'box tree big'. This name is indeed easily analysable as Malyangapa/Wadikali **Thirra-pirna** i.e. *thirra* 'box tree' and *pirna* 'big'.

Figure 20: The landscape north of Yandama Creek, looking from sandhill country south of Gumhole Tank across the claypans and stony plains to the eastern hills. In the foreground an open area with surface exposures of evidence for stone artefact use and production, lies beyond the low shrubs (site HD 2 in McBryde's 1963 survey). Photograph by Isabel McBryde, August 1963. Digitally enhanced by RE Barwick, 2007.

69. 'Dirawanda' (Dutton)

This placename appears as a marginal note in the Beckett manuscript and so it is not clear where in the sequence of sites it comes. The first part of the name is surely *thirra* 'box tree', the second part has several possible etymologies, none of them convincing.

70. 'Balgiri' (Dutton)

George Dutton explained this name as 'Stream waterhole'. Two languages are involved for this area, Pirlatapa and Wadikali. Our information on both is very limited, and so no definite information is available. In Diyari, which is very close to Pirlatapa, *palki-palkiri* means 'wide'.

Figure 21: Stone arrangement, one of twenty similar features on an extensive site (YA 1) recorded by Isabel McBryde in her 1963 archaeological survey of Aboriginal places on Yandama Station. They were associated with surface exposures of stone artefacts. Looking south-east, part of the Three Sisters upland just tops the horizon. Ranging pole marked in feet. Photograph by Isabel McBryde, August 1963. Digitally enhanced by RE Barwick, 2007.

71. 'Winadi Lake' (Dutton)

Modern maps show two names which correspond to 'Winadi':

1. **Winnathee**, the name of a station, about seven kilometres from the border in New South Wales, by the Mundilpa Creek. There is a large swampy area to the southeast of the homestead and this might well have been called 'Winadi Lake' by George Dutton.

2. **Lake Wernathie** is to the northwest, across the Yandama Ck near the Tilcha gate and about 1.5 km on the South Australian side of the border.

It is not clear which of these two sites is involved in the *Ngatyi* story, and it may be that both are involved. Unfortunately, no information is available as to the name.

Note:

> The following two sites are on the Yandama Creek, presumably on the South Australian side.

Figure 22: Remains of the framework of a hut in the sandhill country west of Yandama homestead, where Yandama Creek provided Aboriginal groups with an important route for cultural exchange with the Flinders Ranges. Ranging pole marked in feet. Photograph by Isabel McBryde, August 1963. Digitally enhanced by RE Barwick, 2007.

72. 'Ngamadangga' (Dutton)

All that George Dutton tells us here is 'Ngamadangga – one tree, downstream from Yandama'.

This is a Pirlatapa name, which is also found in Diyari, where it refers to a particular species of tree. It is listed as 'ngamatanka' by Reuther IV no. 3049 in a list of trees and also no. 2597 in a list of 'plants with seeds in pods or capsules', and so it might well be a species of Acacia.[31] Although the word *ngama* means 'milk' over a large area, this plant is not likely to be a Euphorbia with milky sap, as none of that genus grows big enough to be called a tree.

73. 'Gambi gudiandi' (Dutton)

> The Two *Ngatyi* have left their rugs behind at this place: 'So the old fellow said, 'We'll call it Gambi (rugs) gudianda (lost).'

This too is a Pirlatapa name: there is a verb *kuti-* 'to disappear', 'to go away' in the closely related Diyari language. It is used for instance in the sense of rainclouds 'dispersing and disappearing', a starving person's flesh 'dwindling

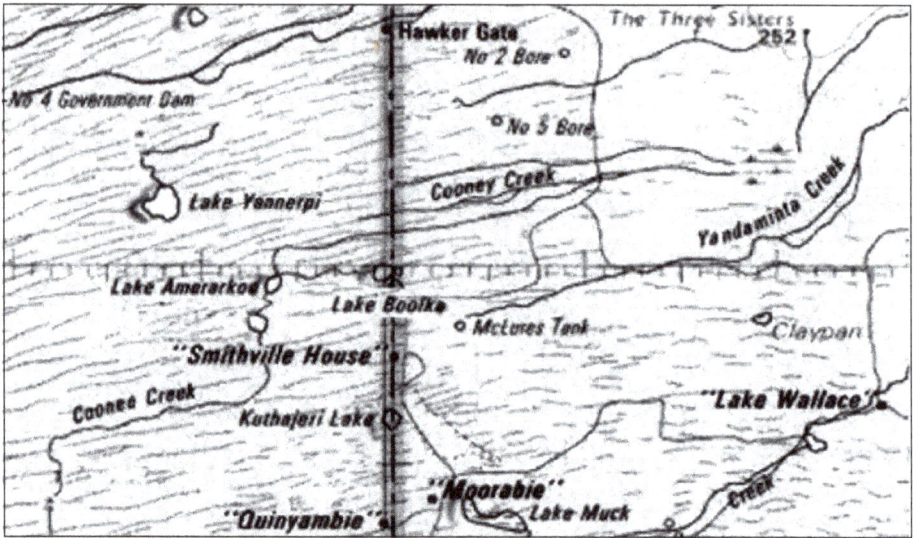

Text Map 2: Lake Boolka

Figure 22a: Cooney Creek. Photograph by John McEntee

and disappearing'.[32] It seems that the verb *kuti-* existed with the same meaning in Pirlatapa though it is not attested in the meagre material that has survived.[33] Hence the name means, more or less as stated by George Dutton, '(our) rugs have gone', (for 'gambi' see no. 62 above).

Nothing is known about a place called 'Angoni' recorded in a marginal note in George Dutton's narrative.

Note:

According to Jeremy Beckett's manuscript and his memories of the occasion George Dutton himself was not clear how all the places in this area fitted in. We have simply listed them in what seemed to be the geographical sequence and have drawn the map accordingly.

74. 'Bulka' (Dutton and Barlow)

George Dutton said, with his usual intellectual honesty: 'When they get down to *Bulka* (beyond Yandama Lake) they travel down through a good bit of country but I don't know the names.' The name 'Bulka', the Boolka of maps, in itself creates some uncertainty. There are seven places near the South Australian–New South Wales border with the name **Boolka**:

1. Boolka Lake on the actual border, 2. Boolka Dam and 3. Boolka Hut adjoining it, 4. Boolka Gate on the border a few kilometres south of the lake, 5. Lake Kamerooka, about 10 km to the south west is called Boolka Lake on the current 1:1 000 000 map. There is also 6. Boolka Soakage on the Coonee Creek some 25 km to the west southwest of the Boolka Gate and there is 7. a Boolka Swamp in the immediate vicinity of that soakage. The place referred to is certainly the first, the lake on the border. In Paakantyi and probably also in this area *pulka* means 'low plain, soggy ground'.

In Alf Barlow's account the *Ngatyi* go from Lake Muck to 'Bulka Lake' and then over the sandhills to Coonee Bore. There is some uncertainty about the details of this name too. The bore is called 'Coonee Creek Bore' on modern maps. The creek on the South Australian side is always spelt 'Coonee' on modern maps and in the South Australian Gazetteer. Only the northern branch of the creek is named in New South Wales and is spelt 'Cooney' (for example Hawker Gate 1:100 000, no 7138 Orthophotomap Series), and on the 1:250 000 maps, as the following extract shows. There is however no difficulty with regard to the location.

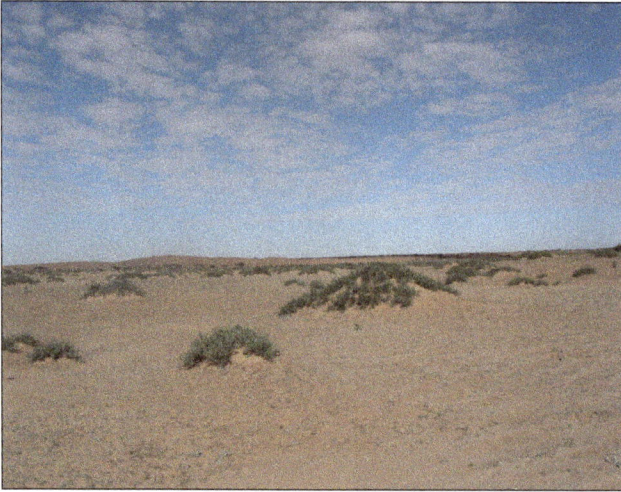

Figure 23: In the sandhills by the lower Yandama Creek. Photograph by Kim McCaul.

Figure 24: Yandama Creek near Yandama Bore. Photograph by Kim McCaul.

75. Coonee Bore

The fact that Coonee Bore is mentioned next makes it even more likely that the site referred to above is indeed no. 1, Boolka Lake on the actual border. George Dutton would probably have followed the same code of naming as Alf Barlow, who also referred to Boolka Lake and Coonee Bore. Coonee represents the word ***kurni*** 'tail' which is found in Malyangapa/Wadikali, Pirlatapa and Diyari. This name is probably connected to a dog myth associated with no. 65 Quinyambie

76. 'Dilarga' (Barlow)

Alf Barlow noted only one site between Boolka Lake and Coonee Bore: and that was Dilarga. He described it as a waterhole presumably in **Coonee Creek,** the only prominent creek in the area. Nothing is known about the name.

Note:

> George Dutton explained that 'down from Yandama Lake' and beyond **Bulka** he did not know the names of the country. This shows that there was a kind of change-over point in this area: Wadikali and Pirlatapa were from there on in charge of the traditions. This change-over point roughly corresponds to the New South Wales–South Australian border. This relative lack of knowledge also reflects George Dutton's own personal life: he had a close association with Yandama, but had not worked on Frome Downs and neighbouring South Australian stations.

As the *Ngatyi* according to all traditions follow the watercourses, we have drawn their track as going down the Yandama-Boolkaree Creek system towards Lake Callabonna. We have no hard evidence for this, but the fact that Alf Barlow's only site on this part of the journey is a waterhole confirms the association with the creek system.

77. 'Durugulili' (Barlow)

The *Ngatyi* here saw that '(a snake) ran away waving about and so made a sandhill 'Durugulili''. This name is analysable as Malyangapa/ Wadikali *thurru* 'snake' and *kuli-* 'play about', 'wander about'.

It is one of the many sandhills between Coonee Creek and the southern part of Lake Callabonna. There is a long particularly wavy one near Lake Wittakilla south of the Yandama Creek, typical of the area, (see Figure 23).

Note:

> George Dutton states: 'Anyway they got about fifty miles on into the **Walpi** and **Biralipa** mob and some of the **Dieri**.' He is here referring to the far northeastern part of the Flinders Ranges. He has mentioned the 'Biralipa' i.e. the Pirlatapa, earlier, and Diyari people were close neighbours to the Pirlatapa in the Blanchwater area.

George Dutton here also mentions the 'Walpi', i.e. **Walypi** or' Wailpi'. This term needs some explanation. The people of the Flinders Ranges now and over the last half century and more identify themselves invariably as *Adnyamathanha* 'the stone people'. This name was known already to Gason 1886 as 'Unyamootha'.[34]

The Diyari similarly called the Flinders Ranges people *Mardala* from the Diyari word *marda* 'stone'. Modern Adnyamathanha i.e. Flinders Ranges people think of the name 'Walypi' as referring *only* to the people who lived around Blinman. The Walypi/ Wailpi are now regarded simply as just a southern part of the Adnyamathanha, and they are described as speaking the same language, though with a special intonation. The late Adnyamathanha speaker Angus McKenzie frequently pointed this out to Luise Hercus: he could mimic the Walypi accent. The Walypi were also said to have had some distinctive and peculiar customs, as is shown in a humorous account by the late Flinders Ranges elder, Andrew Coulthard.[35]

Figure 25: Paralana Creek.
Photograph by Kim McCaul.

Figure 26: Paralana Spring, looking west.
Photograph by Kim McCaul.

**Figure 27: The hot bubbling Paralana Spring.
Photograph by Kim McCaul.**

Tindale on the other hand rejected the use of the name 'Adnyamathanha' as a term for the whole of the Flinders Ranges people and used the term 'Wailpi' instead, giving as his reason 'they are known to the surrounding tribes by that name'.[36]

Tindale continued to use the term, which he wrote as 'Wailpi', to refer to the Flinders Ranges people in general in his papers and in his maps: Elkin also widely used this term. George Dutton's mention of 'Walpi' for people in the far north of the Flinders Ranges is therefore of historical significance: it confirms Tindale's statement about the use of that name by neighbouring tribes instead of 'Adnyamathanha'.

78. Paralana Hot Springs (Dutton)

George Dutton did not name this site but he spoke of 'the waterhole where there's a hot spring' as the place where the *Ngatyi* are made to go back home. Paralana is the major site that fits this description. The name is derived from **Para-ardla-nha** 'dead finish-fire+ suffix *nha*'.[37]

'The two *Ngatyi* 'turned around and said "Alright, wilgamana, but this hole got to be called ngaba galala ganda (water hot)". That's why the water stayed hot.' In Diyari there is a rare verb 'wilka-' which Reuther in a number of entries translates as 'to dig up (wild onions)' and on one occasion, as 'to inflate'.[38] Pirlatapa is very close to Diyari and the chances are that 'wilgamana' is used intransitively with the meaning 'to come out from below': the *Ngatyi,* as one might expect, were deep down in the waterhole and they were saying 'we'll come up (from here)'.

Ngapa is a widespread word for 'water' found in Pirlatapa, Diyari, Wadikali/ Malyangapa and many other languages in the area, but not Adnyamathanha/ Walypi, which has *awi* 'water'. 'Galala ganda' is the Pirlatapa verb form *karlalakarnda* 'stays hot' based on Pirlatapa *karla* 'fire'. The present tense marker is *−arnda*.

79. 'Maliga Lake' (Barlow)

This is one of the most interesting names on the *Ngatyi* track. Despite much searching, we could not find any reference to a place with this name. And then, in June 2003 Luise Hercus went with a small group of people to Paralana Hot Springs and to Lake Callabonna. John McEntee, the owner of Erudina Station and author of an Adnyamathanha (Flinders Ranges) dictionary was a member of this small group.[39] He said: 'In Adnyamathanha *malaka* means "a net bag", that is what Adnyamathanha people used to call Lake Callabonna'. He had this information from Adnyamathanha elders, now deceased. He thought it might be that 'the lake proper is the woven bag, and the narrow channels to the south represent the carrying straps or *yabma*'. In any case the name belonged to the story of the Ancestor *Virdni-murunha* who travelled around the Lake Frome area: some of his story was secret. John McEntee had also found very early evidence of the name *Malaka* for Lake Callabonna in a little known German publication.[40] Further work on this site by the same author is forthcoming and will show that the name has actually survived in an anglicised form to designate other nearby features, both natural and man-made, such as 'Mulligan Springs' and 'Mulligan Hut'.

Note:

It is remarkable that the name 'Maliga', which had apparently been lost from maps, should be remembered over such a distance of space and time by Alf Barlow, whose country lay far to the east. It shows the important role played by *mura*-histories in linking Aboriginal people and traditions.

80. 'Galyamaru' (Dutton), 'Galia maru' (Barlow),'Guliamaru' (Quayle)

Hannah Quayle attached particular significance to 'Guliamaru': in her account the *Ngatyi* died there.

Figure 28: Lake Callabonna. Photograph by Kim McCaul.

Figure 29: The Northern Flinders Ranges from the plain to the east.
Photograph by A-M Siiteri-Hercus.

George Dutton's mention of this site appears as a marginal note in Jeremy Beckett's manuscript at the juncture where the *Ngatyi* turn back from the Hot Springs: '**Galyamaru** Lake between Callabonna and Moolawatana in South Australia'.

By 'Callabonna' he presumably meant the old Callabonna homestead, and Moolawatana would be 'Old Moolawatana'. The only lake roughly in this position is the southern part of Lake Callabonna: this is almost like an appendage to the main lake, and so it is not surprising that it would have had a separate name, 'Galyamaru'.

Alf Barlow gave what turns out to be an indirect explanation of the name when he said:

> *'galia maru'* means 'bottomless'.

ma͟ru means 'black' in all the languages in the immediate area and *(k)alya* means 'loose'. The word **kalya** is pronounced in Adnyamathanha as *alya* because initial *k* is systematically lost in that language, so the name means 'loose black (mud)'. Aboriginal people were very much aware of the treachery of these lake-beds: there is just a thin upper crust and underneath is jet-black ooze which is just as dangerous as any European 'moor': the loose black mud is indeed 'bottomless'. The three storytellers, George Dutton. Alf Barlow and Hannah Quayle all associated this lake with the travels of the *Ngatyi*.

81. Neck of the Salt Creek

This is where the *Ngatyi* were turned back in Alf Barlow's story. He called the site 'Mangunguru'. The interpretation of this name is not known: *mangu* means 'face' in all the languages of the area. Modern maps show 'Salt Creek' to the north of Lake Frome: it forms a narrow neck linking Lake Callabonna to Lake Frome. This is where the waters of Paralana Creek end up: the Paralana creek arises at the Hot Springs, it then flows into the Tea Tree Creek, and this joins the Salt Creek.

The Hot Springs and the Neck of the Salt Creek are not far apart. After travelling many hundreds of kilometres, sometimes on similar tracks, sometimes by widely divergent routes the Two *Ngatyi* in the two main versions of the story turn back in locations that are only about 15 kilometres away from one another. It seems that George Dutton's *Ngatyi* simply travel from the Salt Creek up the Tea Tree Creek to the Paralana Creek, whereas Barlow's *Ngatyi* turn back at the Salt Creek.

The *Ngatyi* travel back the way they had come, according to George Dutton. Alf Barlow's *Ngatyi* however go by a different route on the way back. This takes them past Lake Boolka, closer than they had been on their outward journey to the track of George Dutton's *Ngatyi*. In both versions of the story they do not go to their original home: they go to **Birndiwalpi,** which is envisaged as being linked by deep underground channels to their original home on the Paroo.

82. 'Murlgulu' (Dutton)

As they approach Birndiwalpi the Two *Ngatyi* come to 'a little sandhill there they call *murlgulu* (little bush). When they got there they stood up at each end of the hill. Just down from the sand hill is the water hole *Birndiwalpi*.

The name of the little sandhill is Paakantyi: *murkuru* has been recorded as the word for a small fruit-bearing tree, species uncertain.[41]

83. The Ngatyi get back to Birndiwalpi

In George Dutton's account, the children were at Birndiwalpi: 'when they saw their parents' shadows they started talking in the Malyangapa language'. The first sentence however is in Paaruntyi, the original language of the *Ngatyi*:

In Paaruntyi:

Text	winea	gulbila
Transcription	*wintyika*	*kulypara?*
Translation	Who	shadow?

Who(se) shadow is it?[42]

In Malyangapa:

Text	ingani	nganu	gulbiri
Transcription	*inhanga*	*nganu*	*kulpili*
Translation	Here	my	shadow

It is my shadow.

Text	yaga	ngama bula	duma bula[43]
Transcription	*yakai!*	*ngama-pula*	*kuma-pula*
Translation	Oh	mother-two	father-two

Oh! These are the two, mother and father, father and mother!

Text	mina	wandin dara
Transcription	*minha*	*wanti-rnta-ra*
Translation	what	do (?)-PRESENT-2 plural

What are you all doing?

Text	iba-ni	wirada
Transcription	*ipa-ni*	*wira-tha*
Translation	go in-2sg	hole-into

Go into the hole.

In Alf Barlow's version the young people are singing:

bu:je bu:je **Banmarana**

bu:je bu:je **Baduduru**

jara wala namagulu

gumagula dunggadunggarameida

jardijardi

There are two place-names, **Banmarana** and **Baduduru** in this song. They were not previously mentioned by Alf Barlow, but they are sites mentioned immediately after Birndiwalpi in George Dutton's account of the outward journey (no. 28 and no. 30 above). They are therefore sites via which the two *Ngatyi* would have approached Birndiwalpi on the return journey.

Some words of this song can be analysed as follows:

ngama	mother (with what seems to be an unusual dual marker –*kulu*)
kuma	father (with what seems to be an unusual dual marker –*kulu*)
wala	*warla* 'not'

dunggadunggarameida can be analysed as follows:

final –'eida' represents *(i)tha*, which is the Purposive of a verb in Malyangapa

thungka-thungkarama probably means 'to die', formed from *thungka* 'stinking', 'dead', (cf Galidunggulu, no. 39 above).

wala thungka-thungkaramitha = so that you should not die

The interpretation of 'dunggadunggarameida' is based on a parallel with Paakantyi *puka-malaana* 'to die'. *Puka* is the Paakantyi equivalent of Malyangapa *thungka* and it means 'stinking', 'dead' (as in *Mingka-puka*, no. 10 above).[44]

The meaning of the verse, with some guesswork, would therefore be:

Back to Banmarana

Back to Baduduru

so that mother and father, father and mother should not die.

Note:

The three accounts all converge at this point with the young people singing. This was clearly the focal point of the tradition: here both Barlow and Dutton quote verbatim in Malyangapa the words of the *Ngatyi*. A study of the analysable words of the song and of the conversation raises new questions: 'why are the young ones singing and talking in Malyangapa when they have been on the Paroo all the time?' The answer might be that they are using a language which they assume to be one now more familiar to their parents, who have been travelling further west: though they begin by speaking in Paaruntyi in George Dutton's account, they are talking Malyangapa as a form of politeness.

In all the main versions there is an element of fear at this site. In the Dutton version the parents seem to fear for the young ones and push them back into the hole; in the Barlow version there is certainly a fear of death, but the details are not clear. The Walter Newton version does give us some cause for the fear: 'there was an ordinary big snake – poisonous, just beside the burrows' and it is that big snake which has been singing songs.

In the end, in the main versions, the journey of the *Ngatyi* has run full circle and they are back in the waters of the Paroo, which have come to Birndiwalpi in the underground channels made by the young ones. 'That's how the water got there', said George Dutton.

3.3. Territorial conclusions from the study of the placenames

Of particular interest is the way that the place-names can be attributed to particular languages: the *Ngatyi* speak and name the places in the language of the country they are travelling through. The analysis of placenames fits in exactly with the way George Dutton summarises it: 'they were Paaruntyi first', i.e. they spoke in the Paakantyi language of the Paroo River. When they got to the Yancannia Creek they spoke Pantyikali/Wanyiwalku, the language of the Creek people. From Torowoto Swamp onwards, and in the more favoured area around Yantara Lake there is an overlap, and both Malyangapa and Paakantyi names are used. This was probably jointly owned territory, quite particularly productive after rains when there would have been plenty for everybody. Through the study of the placenames the story of the *Ngatyi* helps to show that 'boundaries' were much more flexible than is often thought; there was ongoing social contact.

On the west side of the Lakes there are still Paakantyi names (nos 46-56). It is only when they get to Mt Brown that the *Ngatyi* 'were 'Wadikalis and Malyangapas'. Beyond 'Mindilba', i.e. Mundilpa Creek, which is a southern branch of the Yandama Creek, they get to 'Birilipa, i.e. Pirlatapa and Wadikali country. As they near the Flinders Ranges they 'run into the Walpi and Biralipa (Pirlatapa) mob and some of the Dieri.'

This corresponds very much to all we know from other sources about the language distribution throughout that large area.

The convergence of the Dutton and Barlow versions is a powerful illustration of the nature of Aboriginal traditions: there is no right version and no wrong version. There are just strong links of storylines that bind the country together across tribal groups, across languages and across many different landscapes.

Endnotes

1. Sutton 2002.

2. Amery 2002: 167.

3. Black forthcoming; McConvell forthcoming.

4. Teulon 1886: 211.

5. Teulon 1886: 211.

6. Jeanette Hope, pers. com.

7. Beckett Jeremy 'The Kangaroo and the Euro, told by George Dutton', unpublished manuscript.

8. Howitt, January 1862, quoted by Shaw 1987: 24.

9. See Shaw 1987: 25.

10. Compare with 'mukko' (Teulon 1886: 211).

11. Teulon 1886: 217.

12. Teulon 1886: 216.

13. Beckler 1993: 63.

14. Tipping 1979: 177.

15. This is probably a misprint for 'places'.

16. Tipping 1979: 176.

17. Tietkens n.d., Reminiscences 1859-87, unpublished manuscript: referring to 1865.

18. Tietkens n.d., Reminiscences 1859-87, unpublished manuscript.

19. McEntee and McKenzie 1992

20. Jack O'Lantern recorded by Luise Hercus May 1985, tape 901

21. Schebeck 2000: 211. Schebeck actually writes 'wida- ukudu', as he uses 'd' to represent the tapped r-sound.

22. Teulon 1886: 212.

23. Richards 1903: 167.

24. Beckett, The Kangaroo and the Euro, told by George Dutton, unpublished manuscript.

25. There is a remote possibility that the second part of this placename is cognate with the Diyari word *'putatja'*, (Reuther 1981, IV: no. 1722) which refers to a small marsupial from whose fur string was woven.

26. Austin 1990.

27. Breen 2004:40.

28. Reuther 1981, VII: no. 929.

29. Mountford, Journals and manuscripts relating to the Flinders Ranges, vol. 10: 129.

30. McEntee and McKenzie 1992: 82.

31. Reuther 1981, IV: no. 3049, no. 2597.

32. Reuther 1981, IV: no. 942.

33. Austin 1990.

34. Gason 1886.

35. Schebeck 1974: 124. Andrew Coulthard describes how long ago, presumably around 1900, his uncle visited Angoorichina near Blinman and was immediately knocked unconscious. The Angoorichina men said: 'this is our custom, our, the Walypis 'custom. This is how we shake hands...

36. Hale and Tindale 1924: 45, quoted by Jones and McEntee 1996: 161.

37. McEntee and McKenzie 1992: 60.

38. Reuther 1981, IV: no. 2608.

39. McEntee 2002.

40. Petermann 1867.

41. Hercus 1993: 46.

42. In Paakantyi and its dialects, as in a number of other Aboriginal languages, the words for 'name' and for 'shadow' are conceived not as 'belonging to oneself', but as actually 'being one's self': that is why they are literally saying 'who shadow this?'

43. This is a misreading for 'guma'.

44. There is a possibility that the *–ma-* suffix here is causative, and that 'wala dunggadunggarameida' might mean 'should not kill (us)'. We think this is unlikely because of the intransitive *–ra* suffix. Paakantyi has *–la,* and in that language *puka-mana* means 'to kill', and *puka-malaana* means 'to die'.

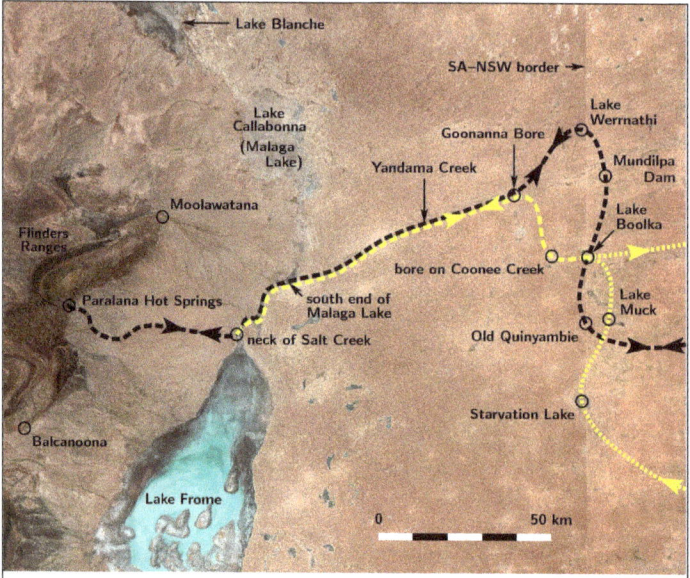

Lake Blanche

SA–NSW border →

Lake Werrnathi

Lake Callabonna (Malaga Lake)

Goonanna Bore

Mundilpa Dam

Yandama Creek

Moolawatana

Lake Boolka

Flinders Ranges

bore on Coonee Creek

Paralana Hot Springs

south end of Malaga Lake

Lake Muck

neck of Salt Creek

Old Quinyambie

Starvation Lake

Balcanoona

Lake Frome

0   50 km

George Dutton's Ngatyi. Part of the route that is traversed in both directions.

Alf Barlow's Ngatyi. Part of the route that is traversed once only.

Alf Barlow's Ngatyi. Part of the route that is traversed in both directions.

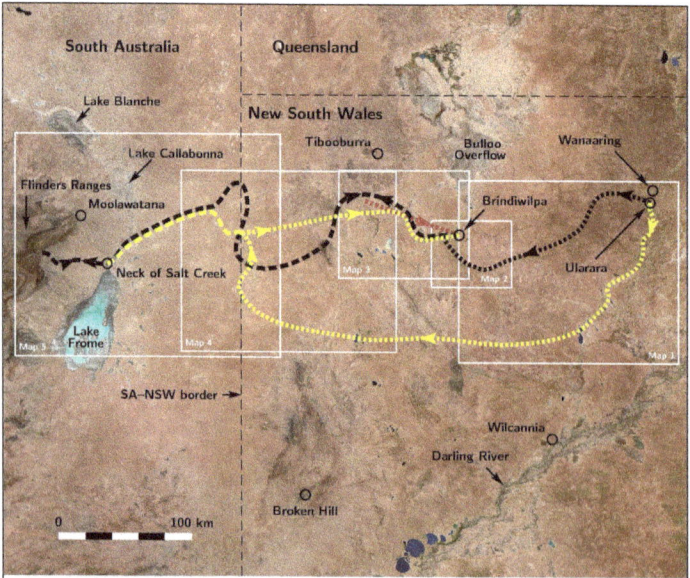

South Australia   Queensland

Lake Blanche

New South Wales

Lake Callabonna

Tibooburra

Bulloo Overflow

Wanaaring

Flinders Ranges

Moolawatana

Brindiwilpa

Neck of Salt Creek

Map 2

Ularara

Lake Frome

Map 5

Map 4

Map 1

SA–NSW border →

Wilcannia

Darling River

Broken Hill

0   100 km

George Dutton's Ngatyi. Part of the route that is traversed once only.

George Dutton's Ngatyi. Part of the route that is traversed in both directions.

Alf Barlow's Ngatyi. Part of the route that is traversed once only.

Alf Barlow's Ngatyi. Part of the route that is traversed in both directions.

Walter Newton's Ngatyi. Route traversed once only.

References **4**

Primary sources

Beckett, Jeremy, The Kangaroo and the Euro, told by George Dutton, unpublished manuscript.

Beckett, Jeremy, Field Notebooks, 1957-8, unpublished manuscript.

Mountford, CP, Journals and manuscripts relating to the Flinders Ranges, vol 10, 'Myths and Material Culture', Mountford-Sheard Collection, State Library of South Australia.

Schebeck, Bernhard n.d., Notes on Kungkatutyi, unpublished manuscript.

Tietkens, WH n.d., Reminiscences 1859-87, unpublished manuscript, State Library of South Australia.

Tindale, Norman B 1938-9, Journal of the Harvard-Adelaide Expedition, unpublished manuscript, South Australian Museum.

Secondary sources

Amery, Rob 2002, 'Weeding out spurious etymologies: Toponyms on the Adelaide plains', in Luise Hercus, Flavia Hodges and Jane Simpson (eds), *The Land is a Map*: *Placenames of Indigenous origin in Australia*, Pandanus Books, Canberra: 165-180.

Austin, AK 1990, 'The last words of Pirlatapa', In *Language and history, Essays in honour of Luise A. Hercus*. Pacific Linguistics Series C, 116, Department of Linguistics, Research School of Pacific Studies, The Australian National University, Canberra: 29-48.

Beckett, Jeremy 1958, 'Marginal men, a study of two half caste Aborigines', *Oceania* 29(2): 91-108.

— 1967, 'Marriage, circumcision and avoidance among the Maljangaba of north-west New South Wales', *Mankind* 6(3): 456-64.

— 1978, 'George Dutton's Country', *Aboriginal History* 2: 2-31.

— 1993, 'Walter Newton's history of the world – or Australia', *American Ethnologist* 20(4): 675-95.

— 1994, 'Aboriginal histories, Aboriginal myths: An introduction', *Oceania* 65(2): 97-115.

Beckler, Hermann 1993, *A journey to Cooper's Creek*, translated and edited by Stephen Jeffries and Michael Kertesz, Melbourne University Press at the Miegunyah Press in association with the State Library of Victoria, Melbourne.

Black, Paul forthcoming, 'Kurtjar Placenames', in HG Koch and LA Hercus (eds), *Aboriginal placenames: naming and re-naming the Australian landscape*.

Blows, Johanna M 1995, *Eagle and Crow: An exploration of an Aboriginal myth*. Garland Pub., New York.

Breen, Gavan 2004, *Innamincka words: Yandruwandha dictionary and stories*, Pacific Linguistics, Canberra.

— 2007, 'Reassessing Karnic', *Australian Journal of Linguistics* 27 (October): 175-99.

Capell, Arthur 1956, *A new approach to Australian linguistics*, Oceania Linguistic Monographs 1, University of Sydney, Sydney.

Curr, Edward M (ed) 1886, *The Australian Race*, vols 1 and 2, Government Printer, Melbourne.

Davis, Stephen 1994, *Australia's extant and imputed traditional Aboriginal territories*, [map], distributed by Melbourne University Press, Melbourne.

Dussart, Françoise 2000, *The politics of ritual in an Aboriginal settlement: Kinship, gender, and the currency of knowledge*, Smithsonian Institution Press, Washington DC.

Elkin, AP 1930, 'The Rainbow-Serpent in Northwest Australia', *Oceania* 1(3): 349-52.

— 1931, 'The social organisation of South Australian tribes', *Oceania* 2(1): 44-73.

— 1938, 'Kinship in South Australia', *Oceania* 8(4): 419-452, 9(1): 41-78.

Gason, S 1886, 'Beltana: Unyamootha Tribe', in Edward M Curr, *The Australian race*, vol 2, Government Printer, Melbourne: 122-3.

Gerritsen, John 1980, *Tibooburra-Corner Country*, Tibooburra Press, printed by Wentworth Press, Sydney.

Hale, HM and NB Tindale 1925, 'Observations on the Aborigines of the Flinders Ranges and records of rock-carvings and paintings', *Records of the South Australian Museum* 3(1): 45-60.

Hardy, Bobbie 1969, *West of the Darling*, Jacaranda, Milton, Queensland.

— 1976, *Lament for the Barkindji The vanished tribes of the Darling River region*, Rigby, Adelaide.

— 1979, *The world owes me nothing: The true story of an Australian Aboriginal's struggle for identity*, Rigby, Adelaide.

Henderson John and Veronica Dobson 1994, *Eastern and Central Arrernte to English dictionary*, Institute for Aboriginal Development, Alice Springs.

Hercus, Luise A 1982, *The Bāgandji Language*, Pacific Linguistics Series B, no 67, Department of Linguistics, Research School of Pacific Studies, The Australian National University, Canberra.

— 1993, *Paakantyi Dictionary*, the author, Canberra.

— and Peter Austin 2004, 'The Yarli languages', in Claire Bowern and Harold Koch (eds), *Australian languages and the comparative method*, Amsterdam studies in the theory and history of linguistic science, Series IV – Current issues in linguistic theory, vol 249, Jon Benjamins Publishing Company, Amsterdam, Philadelphia.

Jones, PG and JC McEntee 1996, 'Aboriginal People of the Flinders Ranges', in M Davies, CR Twidale and MJ Tyler (eds), *Natural history of the Flinders Ranges*, Occasional Publications of the Royal Society of South Australia no 7, Royal Society of South Australia, Adelaide.

Kimber, RG 1986 *Man from Arltunga: Walter Smith, Australian bushman*, Hesperian Press, Victoria Park, Western Australia.

McConnel, Ursula 1930, 'The Rainbow-Serpent in North Queensland', *Oceania* 1(3): 347-9.

McConvell, Patrick forthcoming 'Where the spear sticks up', in HG Koch and LA Hercus (eds), *Aboriginal placenames: naming and re-naming the Australian landscape*.

McEntee, John 1991, 'Lake Frome (South Australia) Aboriginal Trails', *Journal of the Royal Society of South Australia* 115(4):199-205.

— forthcoming, 'An occasion when Mulligan is not just another Irish name, Lake Callabonna, South Australia', in HG Koch and LA Hercus (eds), *Aboriginal placenames: naming and re-naming the Australian landscape*.

— and Pearl McKenzie 1992, *Adnya-math-nha-English dictionary*, the authors, Adelaide.

Mulvaney, DJ 1976, 'The chain of connection', in Nicolas Peterson (ed), *Tribes and boundaries in Australia*, Social Anthropology Series no 10, AIAS, Canberra.

— 1982, *The Aboriginal Photographs of Baldwin Spencer*, introduced by John Mulvaney, Curry O'Neil on behalf of the national Museum of Victoria Council, Melbourne.

Myers, Fred R 1986, *Pintupi country, Pintupi self: sentiment, place, and politics among Western Desert Aborigines*, Australian Institute of Aboriginal Studies and Smithsonian Institution Press, Canberra and Washington DC.

Petermann, A 1867, Mittheilungen aus Justus Perthes Geographischer Anstalt über wichtige neue Erforschungen auf dem Gesammtgebiete der Geographie. By A Petermann. Gotha, Justus Perthes *Geographische Mittheilungen*, Jahrgang 1867 Tafel 18 Vol 13.

Piddington, R 1930, 'The Water-Serpent in Karadjeri Mythology', *Oceania* 1(3): 352-5.

Radcliffe-Brown, AR 1986, 'The Rainbow-Serpent myth in south-east Australia', *Oceania* 1(3): 342-6.

Reid, JA 1886, 'Vocabulary 73: Torrowotto', in Edward M Curr *The Australian race*, vol 2, Government Printer, Melbourne: 178-81.

Reuther, JG 1981, *The Diari*, translated by Phillip A Scherer, Tanunda, South Australia, 1975, AIAS Microfiche no. 2, Australian Institute of Aboriginal Studies, Canberra. (This is a translation of a manuscript from 1904.)

Richards, C 1903, 'The Marra' Waree Tribes or Nation and their language', *Science of Man and Journal of the Royal Anthropological Society of Australasia* 6(8):119-26; 6(11):163-9.

Schebeck, Bernhard 1974, *Texts on the social system of the Atynyamathanha people with grammatical notes*, Pacific Linguistic Series D, no 21, Department of Linguistics, Research School of Pacific Studies, The Australian National University, Canberra.

— 2000, *An Atynyamathanha-English 'research dictionary' version 0.02*, Department of Education, South Australia. Pt Augusta.

Shaw, Mary Turner 1987, *Yancannia Creek*, Melbourne University Press, Carlton.

Sutton, Peter 2002, 'On the translatability of placenames in the Wik region, Cape York Peninsula', in Luise Hercus, Flavia Hodges and Jane Simpson (eds), *The Land is a Map*: *Placenames of Indigenous origin in Australia*, Pandanus Books, Canberra: 75-86.

Teulon, GN 1886, 'Bourke Darling River', in Edward M Curr, *The Australian race*, vol 2, Government Printer, Melbourne: 208-223.

Tindale, NB 1974, *Aboriginal tribes of Australia: Their terrain, environmental controls, distribution, limits, and proper names*, Australian National University Press, Canberra.

Tipping, Marjorie (ed) 1979, Ludwig Becker: Artist & naturalist with the Burke & Wills Expedition, Melbourne University Press for the Library Council of Victoria, Melbourne.

Tunbridge, Dorothy 1988, *Flinders Ranges Dreaming,* Aboriginal Studies Press for the Australian Institute of Aboriginal Studies, Canberra.

Appendix: Transcription of Aboriginal vocabulary 5

When quoting written sources we have left the spelling of words from Aboriginal languages unchanged. When quoting oral sources we have used italics and a practical orthography.

Consonants

The following **consonants** are used in Paakantyi and Malyangapa:

	labial	velar	dental	palatal	alveolar	retroflex
stop	p	k	th	ty	t	rt
nasal	m	ng	nh	ny	n	rn
lateral			lh	ly	l	rl
semivowel	w			y		
r-sounds tap					r	
trill					rr	
glide						r̲

Only these unvoiced consonants have been used in transcription: they are closer to the actual pronunciation, hence we write:

th not 'dh'

t not 'd'

p not 'b'

k not 'g'

There is one exception to this: we have used voiced consonants for the nasal+stop clusters, hence we write *nd, mb*.

Velar nasal

The digraph **ng** is used to transcribe the velar nasal, which is pronounced like the final 'ng' in English 'sing'.